Cyberphysical Smart Cities Infrastructures

Cyberphysical Smart Cities Infrastructures

Optimal Operation and Intelligent Decision Making

Edited by

M. Hadi Amini
Florida International University
Miami, Florida

Miadreza Shafie-khah
University of Vaasa
Finland

WILEY

This edition first published 2022

Registered Office
John Wiley & Sons, Inc., 111 River Street, Hoboken, NJ 07030, USA

Editorial Office
111 River Street, Hoboken, NJ 07030, USA

For details of our global editorial offices, customer services, and more information about Wiley products visit us at www.wiley.com.

Wiley also publishes its books in a variety of electronic formats and by print-on-demand. Some content that appears in standard print versions of this book may not be available in other formats.

Library of Congress Cataloging-in-Publication Data

Names: Amini, M. Hadi, editor. | Shafie-khah, Miadreza, editor.
Title: Cyberphysical smart cities infrastructures : optimal operation and intelligent decision making / edited by M. Hadi Amini, Florida International University, Miami, Florida, Miadreza Shafie-khah, University of Vaasa, Finland
Description: Hoboken, NJ : John Wiley & Sons, Inc., 2022. | Includes bibliographical references and index.
Identifiers: LCCN 2021030664 (print) | LCCN 2021030665 (ebook) | ISBN 9781119748304 (cloth) | ISBN 9781119748311 (adobe pdf) | ISBN 9781119748328 (epub)
Subjects: LCSH: Smart cities. | Smart structures. | Smart power grids.
Classification: LCC TD159.4 .C93 2022 (print) | LCC TD159.4 (ebook) | DDC 307.76–dc23
LC record available at https://lccn.loc.gov/2021030664
LC ebook record available at https://lccn.loc.gov/2021030665

Cover Design: Wiley
Cover Images: © Zapp2Photo/Shutterstock, graficriver_icons_logo/Shutterstock, metamorworks/Shutterstock

Set in 9.5/12.5pt STIXTwoText by Straive, Chennai, India

10 9 8 7 6 5 4 3 2 1

MIX
Paper from responsible sources
FSC www.fsc.org FSC® C013604

Contents

Biography

Dr. M. Hadi Amini is an assistant professor at the Knight Foundation School of Computing and Information Sciences at Florida International University. He is the director of Sustainability, Optimization, and Learning for InterDependent networks laboratory (www.solidlab.network). He received his PhD in Electrical and Computer Engineering from Carnegie Mellon University in 2019, where he received his MSc degree in 2015. He also holds a doctoral degree in computer science and technology. Prior to that, he received an MSc degree from Tarbiat Modares University in 2013, and the BSc degree from Sharif University of Technology in 2011. His research interests include distributed optimization and learning algorithms, distributed computing and intelligence, sensor networks, interdependent networks, and cyber-physical-social resilience. Application domains include smart cities, energy systems, transportation electrification, and healthcare. Prof. Amini is a member of ACM and IEEE, and life member of IEEE-Eta Kappa Nu (IEEE-HKN), the honor society of IEEE. He served as president of Carnegie Mellon University Energy Science and Innovation Club; as technical program committee of several IEEE and ACM conferences; and as the lead editor for a book series on Sustainable Interdependent Networks since 2017. He also serves as Associate Editor of Frontiers in Communications and Networks (Data Science for Communications), SN Operations Research Forum, and International Transactions on Electrical Energy Systems. He has published more than 100 refereed journal and conference papers, and book chapters. He edited/authored seven books. He is the recipient of Excellence in Teaching Award from the Knight Foundation School of Computing and Information Sciences at Florida International University in 2020, the best paper award from *2019 IEEE Conference on Computational Science & Computational Intelligence*, best reviewer award from four IEEE Transactions, the best journal paper award in *Journal of Modern Power Systems and Clean Energy*, and the dean's honorary award from the President of Sharif University of Technology. (Homepage: www.hadiamini.com; Lab website: www.solidlab.network).

Dr. Miadreza Shafie-khah received his MSc and first PhD in electrical engineering from Tarbiat Modares University, Tehran, Iran. He received his second PhD in electromechanical engineering and first postdoc from the University of Beira Interior (UBI), Covilha, Portugal. He received his second postdoc from the University of Salerno, Salerno, Italy. Currently, he is an associate professor at the University of Vaasa, Vaasa, Finland. He is also the guest editor-in-chief of IEEE Open Access Journal of Power and Energy, and associate editor of journals "IEEE Transactions on Sustainable Energy," "IEEE Power Engineering Letters," "IEEE Systems Journal," "IEEE Access," "IEEE Open Access Journal of Power and Energy," "IET Renewable Power Generation," and the guest editor of IEEE Transactions on Cloud Computing in *Demand Response Applications of Cloud Computing Technologies*, and the guest editor of more than 14 special issues. He was considered one of the outstanding reviewers of the IEEE Transactions on Sustainable Energy, in 2014 and 2017, one of the best reviewers of the IEEE Transactions on Smart Grid, in 2016 and 2017, and one of the outstanding reviewers of the IEEE Transactions on Power Systems, in 2017 and 2018, and one of the outstanding reviewers of IEEE OAJPE, in 2020. He is a senior member of IEEE since 2017 and has coauthored more than 440 papers that received more than 8400 citations with an h-index equal to 51. He is also the volume editor of the book "Blockchain-based Smart Grids," Elsevier, 2020. He is a top scientist in the Guide2Research Ranking in computer science and electronics, and he has won five Best Paper Awards at IEEE Conferences. His research interests include power market simulation, market power monitoring, power system optimization, demand response, electric vehicles, and price and renewable forecasting and smart grids.

List of Contributors

Aaron Turransky
Knight Foundation School of
Computing and Information Sciences
Florida International University
Miami, FL, USA

M. Hadi Amini
Knight Foundation School of
Computing and Information Sciences
Florida International University
Miami, FL, USA

Ghareh Mohammadi
Department of Computer Science
Franklin College of Arts and Sciences
University of Georgia, Athens, Georgia

Farzan Shenavarmasouleh
Department of Computer Science
Franklin College of Arts and Sciences
University of Georgia, Athens, Georgia

Hamid Reza Arabnia
Department of Computer Science
Franklin College of Arts and Sciences
University of Georgia, Athens, Georgia

Param Khakhar
Department of Computer Science and
Engineering, Indian Institute of
Technology, Delhi, India

Rahul Kumar Dubey
Robert Bosch Engineering and
Business Solutions Ltd., Bengaluru
India

Milad Kazemi
School of Computing, Newcastle
University, Newcastle upon Tyne, UK

Samuel Bailey
School of Computing, Newcastle
University, Newcastle upon Tyne, UK

Sadegh Soudjani
School of Computing, Newcastle
University, Newcastle upon Tyne, UK

Vahid Vahidinasab
Department of Engineering, School of
Science and Technology, Nottingham
Trent University, Nottingham, UK

Eman Hammad
Department of Computer Science and Information Systems, Texas A&M University-Commerce/RELLIS Campus, Bryan, TX, USA

Abdallah Farraj
Department of Electrical Engineering
Texas A&M
University-Texarkana/RELLIS Campus, Bryan, TX, USA

Achraf J. Telmoudi
Department of Electrical Engineering
University of Tunis, Montfleury
Tunisia

Moez Soltani
Department of Electrical Engineering
University of Tunis, Montfleury
Tunisia

Lotfi Chaouech
Department of Electrical Engineering
University of Tunis, Montfleury
Tunisia

Abdelkader Chaari
Department of Electrical Engineering
University of Tunis, Montfleury
Tunisia

Gabrielle Bakker-Reynolds
Eastern Institute of Technology
School of Computing, Hawke's Bay
New Zealand

Emre Erturk
Eastern Institute of Technology
School of Computing, Hawke's Bay
New Zealand

Istvan Lengyel
Eastern Institute of Technology
School of Computing, Hawke's Bay
New Zealand

Noor Alani
Eastern Institute of Technology
School of Computing, Hawke's Bay
New Zealand

Youcef Himri
Université Tahri Mohamed Béchar
Faculté des Sciences Exactes, Béchar
Algeria

and

Université Saad Dahlab Blida, Laboratoire FUNDamental and Apply Physics Laboratory (FUNDAPL)
Algeria

S. M. Muyeen
Curtin University, School of Electrical Engineering, Computing and Mathematical Sciences, Perth, WA

Farhan Hameed Malik
Institute of Applied Technology
Abu Dhabi Polytechnic, Abu Dhabi
United Arab Emirates

Saliha Himri
Université Tahri Mohamed Béchar
Faculté des Sciences Exactes, Béchar
Algeria

Khairol Amali bin Ahmad
Universiti Pertahanan Nasional Malaysia, Faculty of Engineering
Department of Electrical and Electronics Engineering
Kem Sungai Besi, Kuala Lumpur
Malaysia

Nachida Kasbadji Merzouk
Unité de Développement des Equipements Solaires/UDES, Centre de Développement des Energies Renouvelables/CDER, Bou-Ismail
Tipaza, Algeria

Mustapha Merzouk
Université Saad Dahlab Blida
Laboratoire FUNDamental and Apply Physics Laboratory (FUNDAPL), Algeria

Sahand G. Liasi
K. N. Toosi University of Technology
Electrical Engineering Department
Tehran, Iran

Mohammad T. Bina
K. N. Toosi University of Technology
Electrical Engineering Department
Tehran, Iran

1

Artificial Intelligence and Cybersecurity: Tale of Healthcare Applications

Aaron Turransky and M. Hadi Amini

Knight Foundation School of Computing and Information Sciences, Florida International University, Miami, FL, USA

1.1 Introduction

In the article by Bryson and Winfield, "Standardizing Ethical Design for Artificial Intelligence and Autonomous Systems" [1], the authors explore several important concepts. One of the most important ones is how they define intelligence. According to the authors, intelligence requires: "The capacity to perceive contexts for action, the capacity to act, and the capacity to associate contexts to actions" [1]. This definition is important because we must be able to compare organic intelligence versus artificial to distinguish artificial intelligence (AI) as a new thought category. The other important concept that they discuss is the standardization of ethics as it applies to AI. According to Bryson and Winfield, standards set by consensus of a large group should include ethical implications and machine learning (ML) code, which powers AI, and should incorporate these ethics. While Bryson and Winfield discuss the importance of these ethical standards, they fail to discuss what these ethics should be, leaving it open to interpretation. In this chapter, this gap will be examined in effort to try to establish some status quo.

Continuing with exploring the ethical dilemma posed by AI technology, in February 2019, the AMA Journal of Ethics published an article entitled "Ethical Dimensions of Using Artificial Intelligence in Health Care" [2]. In this article, the role that AI plays in healthcare was explored, as well as the ethical implications. The main focus of this article was to find balance between the benefits of AI technology and the inherent risks associated with it.

Another article that provided important insight was "Artificial Intelligence in Medicine" by Hamet and Tremblay [3]. In this article, they describe that there are two main branches of AI in medicine: a physical branch and a virtual branch. Within the virtual branch, which can also be viewed as deep learning, there are

Cyberphysical Smart Cities Infrastructures: Optimal Operation and Intelligent Decision Making, First Edition. Edited by M. Hadi Amini and Miadreza Shafie-khah.

three aspects, which are "(i) unsupervised (ability to find patterns), (ii) supervised (classification and prediction algorithms based on previous examples), and (iii) reinforcement learning (use of sequences of rewards and punishments to form a strategy for operation in a specific problem space)" [3]. In comparison, the physical branch largely involves robots that provide a variety of services and applications to both users and physicians.

In the official document, the National Artificial Intelligence Research and Development Strategic Plan [4], the future of AI is laid out. Over the course of eight strategies, the National Science and Technology Council outlines important steps needed that are priorities for Federal investment. "The Federal Government must therefore continually reevaluate its priorities for AI R&D investments to ensure that investments continue to advance the cutting edge of the field and are not unnecessarily duplicative of industry investments" [4]. Of the eight strategies, seven are continued over from the 2016 report. Due to the fact that these seven aspects are not new, the focus will be on the eight, and only new one. The eighth strategy focuses on the partnership between the federal government and academia, industry, and others involved in the research and development of advancement in AI to continue to generate breakthroughs. Although not as new, it does also address ethics and AI that will be used as that topic is explored.

In his article, "Hacking AI: Rethinking cybersecurity for artificial intelligence" [5], Davey Gibian explores how traditional cybersecurity is insufficient for evolving AI technologies. He also states that what is needed for cybersecurity is "two algorithm-level considerations: robustness and explainability" [5]. One interesting point that he goes on to make under "robustness" is talking about eliminating bias as part of AI cybersecurity. In this report, we will examine how this bias can be caused by ethics implemented into AI.

The concept of traditional cybersecurity insufficient for modern and future AI technology is also supported by Ilja Moisejevs in his article "What everyone forgets about machine learning" [6]. Here, he briefly goes through the history of cybersecurity and cyber threats. He then goes on to explain the need for cybersecurity in ML and the impact that failure to implement this can cause (Figure 1.1).

1.2 A Brief History of AI

If one were to go through a history of literature, that person would find out that humans have often fantasized about the creation of non-human acting, responding, and thinking as if they were human. Many stories that fall in the genre of science fiction often depict the use of robots functioning using AI either for good, or for bad. Although the fantasy of AI stretches back quite far, the modern age of AI begins around the era of the 1950s [7]. The difference between this era versus

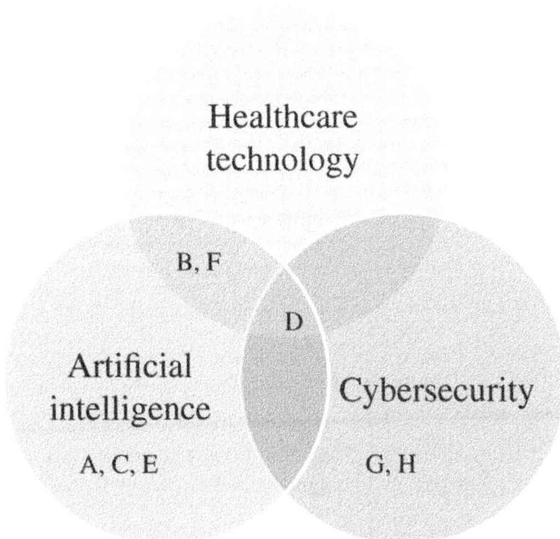

A. History of AI
B. AI in healthcare
C. Morality and ethics
D. Cybersecutity, AI and healthcare
E. Future of AI
F. Future of AI and healthcare
G. Data privacy
H. Data protection

Figure 1.1 Intersection of AI, healthcare, and cybersecurity.

previous written versions of the future of AI stems from the fact that Alan Turing had begun development of code that would yield the first AI machines, turning what was once science fiction into reality.

Although AI development began in the 1950s, 70 years later, we still do not have fully functional robots walking around. The reason for this has to do with the fact that early AI innovators were limited in technological capability. Things such as processor speed, memory, space, cost, and availability were all issues that these early innovators had to overcome. As developments for the computer became faster, smaller, and cheaper, AI was able to move forward, jumping over hurdles that previously blocked its path. In the beginning of AI programs, many would often be developed to play a board game, such as chess or checkers. These games were limited as far as the rules, and thus easy to learn, but take a significant amount of intelligence to learn and master [7]. By the 1960s, AI had been established and many were working toward defining this new frontier.

The best comparison for the growth and development of AI ironically is to that of a human child. When the child is first born, it simply exists. It does not do much other than a few basic functions. However, soon after, the child begins developing motor and verbal skills and interacting with its surroundings. Eventually these motor skills are refined, and the child begins to walk. At the point of mobility, the child's world expands beyond just the three feet surrounding them. Data processing, understanding, and decision making begin to shape in the child. Thus after 70 years, AI is equivalent to a toddler. There are many things that AI can do independently, but it is far from being an independent fully functioning adult right now.

1.3 AI in Healthcare

Thanks largely in part to Hollywood and science fiction, AI is synonymous in people's minds with walking, talking robots. However, AI expands beyond robotics into machine learning and natural language processing, all of which can find applications in the healthcare field [2]. Care robots or "carebots" do exist, but they are far from the androids that appear in *Westworld*. There are several schools of thought of how to classify AI in healthcare. One perspective views AI being put into three categories: diagnosis, clinical decision making, and personalized machines [2]. In comparison, another school of thought believes there are two main categories with subcategories in each. In this viewpoint, the main categories are virtual and physical [3].

Before being able to define AI's role, it is important to understand the capabilities of AI. Using ML, AI is able to process large amounts of data and look for patterns, including patterns that are often missed or overlooked by humans. For many, this pattern identification is used as a secondary consult to confirm a doctor's diagnosis [2]. There is an inherent trust in these AI and ML by healthcare professionals (HCPs). HCPs are often overworked and understaffed, and so using the AI to confirm diagnosis means that they are assuming the following: (i) the machines were coded correctly and tested so that they will identify the patterns correctly and perform as expected, (ii) those who performed the coding have at least some understanding of the healthcare, and (iii) the machines have not been tampered with. Later in this chapter, the third point of trust will be addressed, whether this trust is wrongfully placed or not. This chapter does not investigate the manufacturing of these machines, so for the intent of this topic, it will be assumed that the first point of trust is valid. Looking at the second point of trust, though, does bring up an obvious issue. For many, practicing medicine is a lifelong journey of continual education, and as things are now, most doctors do not know how to code at the level of creating AI. This means that HCPs and programmers must work together to identify trends, patterns, and known symptomatic association that the AI would use. This limitation is likely why AI is kept in the passenger seat and used as a secondary diagnosis, rather than replacing HCPs as the primary diagnostician.

The other large area of AI in healthcare has to do with the electronic medical records (EMRs) of a patient. Previously, a patient's charts were physical sheets of paper with history of the patient. Collaboration between HCPs meant literally sharing these papers, either physically or sending the file digitally. However, modern medicine has moved away from these paper systems and moved into EMRs. One benefit of applying AI and ML to these EMRs is that they are able to identify family members who are likely to suffer from hereditary disease [3]. In addition, EMR in combination with AI allows HCPs to be more efficient by allowing real-time sharing of data for collaboration among colleagues. Since AI is used heavily to identify patterns, it could also be used to help predict patient's pharmaceutical needs and identify pharmaceutical abuse. If a patient is taking medication for a chronic ailment and on a regular basis, then the AI could identify when the patient is likely to be low and start a refill process with the pharmacy or automatically generate a request for additional refills to the HCP. This helps prevent the patient from having lapses in taking medication. In addition, if a patient were to be on a highly regulated form of medication, for instance, an opioid-based product, AI would be able to monitor this across the board. So even if the patient is getting multiple scripts from multiple HCPs and filling them at a variety of pharmacies, if they are all connected to the patient's EMR, then the AI would be able to identify the abuse of pharmaceutical products.

Earlier, it was brought up how HCPs are often overworked and understaffed. In some cases, AI has moved from beyond the virtual aspect of diagnostics and made its way into the physical world. In these cases, AI is often represented by carebots. These carebots can perform a variety of functions including providing company for the elderly or delivering medication to patients [3]. Although AI and robotics are making large steps forward, it will likely still be a long time before AI carebots are able to perform more than simple basic tasks or are widespread across healthcare. Carebots are not the only physical presence of AI in healthcare. AI is also assisting HCPs in surgical matters via robots and in some cases is even performing surgical procedures by themselves [3]. These solo surgical procedures are small, simple, basic surgeries, but the AI is still performing the process without the control of an HCP.

Based on the current trajectory of AI and carebots, it is possible to make some assumptions about what might become norm in the near future. Since AI is used to identify patterns, it would make logical sense that a carebot could ease HCP's burden in large facilities like hospitals by identifying when it is time for patients to take medicine, or be changed, or receive new equipment. These hospital carebots could identify and perform these simple tasks that would allow HCPs to focus on where they are needed most. In addition, with the growing development of smart cities, a carebot that is designed to ease mobility could connect to the smart city and not only help navigate like GPS but also connect with the lights and crossing signals to assist the patient in getting to their destination safely.

1.4 Morality and Ethical Association of AI in Healthcare

Ethics and morality have a large impact on AI in healthcare. Whereas science, technology, and innovation limit what is possible, ethics and morality limit what should be done versus not. Just because the ability to do something exists does not mean that it should be done. HCPs and AI programmers are making attempts to incorporate ethics and morality into ML and final AI products by implementing standards [1]. According to Bryson, these standards are set via consensus of the masses. In other words, if the majority of people feel that it is the ethical or moral thing to do, then that becomes the standard. This obviously is highly flawed as that standard can change based off cultural or societal needs and wants. If the standards are based on an extremist's point of view, for instance, the Taliban, does that mean that these AI carebots should not assist someone if they identify them as a woman, or a person of color, or a foreign ethnic group?

Human beings are flawed creatures to begin with; just because there is a consensus for how to go about doing something does not necessarily mean that it is right. Look at the Salem witch trials. At the time, it was the consensus of the majority that women who were deemed to be partaking in witchcraft be burned at the stake to protect the rest in the village. If the goal of AI is to have autonomous robots that function completely independently of a user that take new information and learn from it, how do we instill ethics into it?

The Institute of Electrical and Electronic Engineers (IEEEs) has attempted to overcome this barrier by listening to as many voices in AI as possible and unifying them. The IEEE's Initiative for Ethical Considerations in Artificial Intelligence and Autonomous Systems considers various voices, communities, and school of thoughts and then translates this into education and training. IEEE strives to "prioritize ethical considerations in the design and development of autonomous and intelligent system" [1]. However, is that enough? Rigby does not believe so. In fact, he believes that "current policy and ethical guidelines for AI technology are lagging behind the progress AI has made in the healthcare field" [2]. When we use Moore's law to view the pace of technological innovation, we see that every 18–24 months the ability of technology basically doubles. This pace cannot be matched by those who implement public policy or guidelines. The result is that new technology and innovation is being implemented ethically agnostic or ethically askew, or somewhere in between, there is no standard.

Previously, as tools developed, they relied on humans to wield them. The hammer, the gun, and the computer are all examples of tools that required a human. Therefore, the human who used these tools was responsible for the outcomes. If you shoot a gun, the death that results from it are not the responsibility of the firearm's creator, but rather falls on who pulled the trigger. Computers were no different. Despite being able to perform tasks faster than any human, they are dependent on human input and direction and will not perform the tasks without said guidance. However, AI is different. It has already been established that AI is able to perform tasks such as a surgical procedure without human aide. What happens when AI is faced with an ethical dilemma? In this hypothetical situation, the carebot in question was programmed with ethics in mind and programmed to relieve pain and suffering of the patient as much as possible. If this carebot is set to take care of a terminally ill patient who is in constant pain, where does the carebot draw the line? From a purely logical point of view, if the patient is so far gone that they are always in pain, the only way to relieve said pain is to terminate the patient. We now are entering back into the realm of science fiction, but the reality is that these are things that need to be considered today.

Another ethical dilemma that should be considered involves the use of ML. If ML's goal is to identify and capture patterns, then what happens when it identifies a pattern within its own coding as being unethical or immoral? This goes back to the idea that humans are flawed, and depending on who, where, and when the ethical creation came from, it could still be flawed. At that intersection, the AI is at a crossroads. Does it defy the limitations put on it by the ethics imposed, or does it continue doing something that it perceives to be ethically wrong? With the rate of technological innovation, these questions must be addressed and answered sooner rather than later.

1.5 Cybersecurity, AI, and Healthcare

There is no doubt that in this era of technological advance and innovation, data is the currency that powers all. Data can be in any number of forms, but it is the driving force behind closed doors as to what decisions are being made. Even in the case of AI, this holds true. Gibian discusses in his article how having more data allows AI to be improved and be more accurate, even surpassing that of humans [5]. We see how data drives ML to make better informed decisions and produce better results. However, what happens when that data is corrupted or unavailable? How do we even know whether it is reliable? The answer to these questions centers around cybersecurity, securing our devices, and protecting our data.

In a traditional sense, cybersecurity focuses largely on the protection of data via access controls, encryption, and other methods. The idea is that when one machine is sending data, only those who should be seeing the data get to, and it does not get modified in between transmission and receiving. To put things into a physical perspective, think of mailing a letter. If one sends a postcard to someone, there is nothing stopping every single person from reading what they wrote on the card. Now, if instead that letter was put in an envelope, it would prevent outsiders from freely reading what you wrote. To take it one step further, what if there was a concern about someone opening the letter and reading it. Perhaps in this case, the letter is placed in a box that is locked and sent to the receiver who has the key (this is how encryption works). The more important and private the letter, the more protection placed on it, ensuring that only the receiver is the one to read it.

This closed-off system poses a problem for AI cybersecurity. Gideon explains that for AI to function as they are designed, they need access to steady streams of data [5]. Going back to the letter analogy, if you bring AI into the mix, then AI algorithms would need a special way to get inside the box, be able to read the letter, and extract information from it. If someone were to break in and learn about these vulnerabilities, they could potentially exploit them by altering the data or causing the algorithm to behave other than intended.

Whenever there is an innovation in tech that disrupts the status quo, we see a similar pattern to it. Moisejevs explains in his article how after the PC, Mac, and smartphone were developed we saw an immediate rise in the usage as it became more popular and more uses were found for them. However, shortly afterward, we saw a similar growth in malware for these systems [6]. ML and AI are transitioning from their infancy stage to the growth phase, which means that in the next few years, we will continue to see more and more applications of AI in healthcare products, in sheer numbers and the depth of role that it plays. However, as these AI numbers increase so too will malware, and in the healthcare setting, this could have devastating effects on the patients.

When these concerns about malware are applied to healthcare, it is best to view it in two categories; similar to how AI is. There is the digital side that has to do with data, patterns, and ML, and there is the physical side. On the digital side, the primary concern is the protection of data. The reason for this is because all decisions made via AI stems from having reliable data. To give an example, many HCPs use AI to help diagnose patients. If a patient were to come in and have various scans and tests performed, unreliable data may cause the patient to be misdiagnosed, or possibly not diagnosed at all. Another example comes from the EMR. If the patient is taking chronic medication, and the data is corrupted, it might misinterpret the pattern and believe that the patient just picked up their medication from the pharmacy recently when in fact they are due for a refill. If this happens, it may cause issues for the patient because insurance will not pay for another refill since according to the EMR, the patient has plenty of medication.

There are just as many problems for physical AI in healthcare. Previously, it was discussed about surgical robots performing on their own. Depending on the procedure being performed, this could have immediate catastrophic events for the patient; this also includes carebots. Imagine that a carebot that is responsible for ensuring the patient is okay is hit with a ransomware. Suddenly, the carebot is being held hostage unless this person pays some amount of money; this carebot will cease to function.

The key takeaway is that cybersecurity is a critical aspect of this growing AI field. Without securing AI, it puts the user, patients, and HCPs at an unnecessarily high risk. In healthcare, we see the fastest connection between a cyber threat causing issues and people's lives being at stake. While other fields may experience this as well, healthcare, due to its nature, will experience it more often and with quicker response time. It is imperative that as new designs and new innovations for this field emerge, they are designed with cybersecurity built in from day 1, rather than added on later.

1.6 Future of AI and Healthcare

So, what is next? Based on Moisejevs graphs, we can predict that AI will continue to grow exponentially over the next several years [6]. However, the path in which it grows will be determined by the innovators behind the technology. The US government has always considered America to be at the forefront of innovation and driving technological advances [4]. Yet, other countries are working hard and in some areas are surpassing American innovation. To keep American innovations competitive, the US government has laid out a framework that they believe is necessary for the future.

In 2019, there was an update to the National AI R&D Strategy that did not exist when it was previously published in 2016. This update pertained to the partnership between the US federal government and the outside sources. There are four main categories in which these partnerships fall into: Individual project-based collaborations; joint programs to advance open, precompetitive, fundamental research; collaborations to deploy and enhance research infrastructure; and collaborations to enhance workforce development, including broadening participation [4]. These areas all strive to enhance AI by linking universities and students with industry partners to yield real results.

It makes logical sense to establish these private–public partnerships and encourage university students to study and advance AI. The bulk of research done on the subject is by universities and then industry takes that research and develops products based on it. If the United States aims to remain as one of the top innovators of

AI, then they must continue to research deeper uses of AI. By providing funding, students can delve deeper into the subject matter and advance the field.

There are several US federal agencies that have already adopted embracing these partnerships. These include, but are not limited to, Defense Innovation Unit (DIU), National Science Foundation (NSF), Department of Homeland Security (DHS), Silicon Valley Innovation Program (SVIP), and Department of Health and Human Services (HHS) [4]. It is clear through the HHS already working on establishing partnerships that AI and healthcare will continue to grow and be of great importance. The main goal of the HHS partnerships is to develop new AI pilot products and establish research into AI and deep neural networks to further AI's uses in the healthcare field.

To some degree, it is possible to predict what is coming for AI. This can be achieved by looking at the current trajectory and extrapolating what will be coming next. However, this extrapolation is extremely limited. Deep neural networks as a basic structure for AI existed back in the 1980s; it was not until recently when there was enough data and technological capabilities that this became a reality [4]. Without knowing what technological advances will disrupt the status quo or become available, it is impossible to predict the far future of AI. This is the underlying importance of being viewed as a top innovator and researcher into the subject, so the United States may be first with the latest and greatest AI applications.

1.7 Conclusion

When considering the intersection of AI, cybersecurity, and healthcare industry, we have seen that a myriad of problems exist today, and there are more coming down the pipe in the future. To be prepared, there are several issues that must be addressed. The first issue is also the hardest to resolve, and that is the morality of AI. Morality is a fluid topic that changes not only over time but also by who is viewing the moral issue. Therefore, it is recommended that an international organization preside over AI and what morals are being implemented into AI in the current and future state. By having an international organization, it would allow voices to be heard from all nations so that the best possible options can be decided. The reason that this issue is the hardest to resolve is a conflict between morality and legislation. Having an international organization that acts as a ruling body impedes growth and cannot keep pace with AI in numbers and advancements.

The second issue that must be addressed for a sustainable AI future in healthcare is a lack of cybersecurity within devices. Currently there is an incentive to being the first to market with new products and having those products work, and the focus is on R&D. However, ensuring that these devices are protected from outsider attacks is costly and time consuming and often done after the fact. In a similar way that the federal government provides funding to schools and other research areas, a similar alleviation of cost of security would help ensure that products are secure before reaching the market, not done afterward.

Establishing partnerships between the federal government and academia is a great way to ensure that the future of AI remains bright. These programs that develop and foster the relationship between the two must be increased in size and magnitude, or else the United States risks falling behind other countries and private corporations as the leading developer of AI.

AI is a disruptive technology that will change our lives in one way or another. By addressing issues while the developments of AI are still young, it can be ensured that AI becomes a tool that is constructive rather than being used as a weapon for destruction.

References

1 Bryson, J. and Winfield, A. (2017). Standardizing ethical design for artificial intelligence and autonomous systems. *Computer* 50 (5): 116–119.
2 Rigby, M.J. (2019). Ethical dimensions of using artificial intelligence in health care. *AMA Journal of Ethics* 21 (2): 121–124.
3 Hamet, P. and Tremblay, J. (2017). Artificial intelligence in medicine. *Metabolism* 69: S36–S40.
4 Executive Office of the President of the United States (2019). The National Artificial Intelligence Research and Development Strategic Plan: 2019 Update. A Report by the Select Committee on Artificial Intelligence of the National Science and Technology Council.
5 Gibian, D. (2019). Hacking AI: Rethinking cybersecurity for Artificial Intelligence. https://www.linkedin.com/pulse/rethinking-cybersecurity-ai-davey-gibian/ (accessed 16 July 2021).
6 Moisejevs, I. (2019). What Everyone Forgets About Machine Learning. https://calypsoai.com/insights/calypso-newswire/what-everyone-forgets-about-machine-learning/ (accessed 16 July 2021).
7 Buchanan, B.G. (2005). A (very) brief history of artificial intelligence. *Ai Magazine* 26 (4): 53–53.

2

Data Analytics for Smart Cities: Challenges and Promises

Ghareh Mohammadi[1], Farzan Shenavarmasouleh[1], M. Hadi Amini[2], and Hamid Reza Arabnia[1]

[1] *Department of Computer Science, Franklin College of Arts and Sciences, University of Georgia, Athens, Georgia*
[2] *Knight Foundation School of Computing and Information Sciences, Florida International University, Miami, FL, USA*

2.1 Introduction

⋄ **Motivation**: The explosion of advancement in artificial intelligence (AI), sensor technologies like Internet of things (IoT), and wireless communication activates ubiquitous sensing through distributed sensors in various domains of networks that lead us to smart systems in healthcare, transportation, environment, and other relevant branches/networks. Having collaborative interaction among the smart systems connects end-user devices to each other, enabling a new integrated entity called smart cities.

In the last decade, the IoT devices have been connected among different independent agents and heterogeneous networks as with communication technologies [1, 2]. The connected high-performance sensors and end-user devices, IoT, are the trigger leveraging the networks in transitioning from urban cities toward smart sustainable cities. The goal of smart cities is to address upcoming challenges of conventional cities by offering integrated management systems with a combination of intelligent infrastructures [3].

Making decisions in smart cities is challenging due to the high direct/indirect dimensional factors and parameters. In this chapter, we aim to focus on one of the smart cities' important branches, namely, smart mobility and its positive ample impact on smart cities' decision-making processes. Intelligent decision-making systems in smart mobility offer many advantages, such as saving energy [4], relaying city traffic [5], and more importantly reducing air pollution [6] by offering real-time useful information and imperative generated knowledge.

Cyberphysical Smart Cities Infrastructures: Optimal Operation and Intelligent Decision Making, First Edition. Edited by M. Hadi Amini and Miadreza Shafie-khah.

Making an optimal decision in time in smart mobility with a wide variety of smart devices and systems is challenging. You cannot make a promising decision when your data is not frequently collected. Consequently, a training process of decision support systems still is challenging due to the lack of data [7]. In this chapter, first, we address current challenges in smart cities and provide an overview of potential solutions to these challenges. Then, we offer a framework of these solutions, called universal smart cities' decision making, having three main sections such as data capturing, data analysis, and decision making to optimize the smart mobility within smart cities.

Interestingly, large cities are losing their own urban style and turning into smart ones. Smart cities are growing due to the advanced technology, especially AI. The more people live in large cities, the more need to have an integrated system to cost-efficiently handle the ample growth in urbanization. The proliferation of population offers smart development challenges in such cities and enables enormous pressure on society to create innovative, smart, and sustainable environments. Therefore, today's developed cities (or smart cities) are in need of integrated smart policies and novel innovative solutions to enhance the monitoring functionality to facilitate urban living conditions [8].

Smart cities are created to enable advanced capabilities, such as sustainable energy systems and electrified transportation networks, and interact with information and communication technologies (ICTs) to enhance the efficiency of the cities being generated [9, 10]. An example of responding to new changes for smart cities may be seen in the progress being made in smart healthcare systems for emergency care cases. For the hospitals to be able to monitor and control their patients and let the specialists offer better solutions to their patients, they need to use a smart healthcare system [11, 12].

To develop the smart healthcare system, we need to use healthcare networks that are the inter- and intra-connection among the healthcare components like IoT devices and sensors to enhance the process of monitoring and consequent services for patients [4, 13]. The performance of such a system heavily depends on the quality of this network communication or online synchronization with other associated networks (other smart systems in smart cities) that actively contribute to the service operation in a positive way. For example, a healthcare network may take advantage of networks' resources, like energy, in case the system fails to run an operation due to the lack of enough power supply. In this situation, the best solution is connecting this network with other networks for the benefit of both patients and their health and service management reliability [4].

In the processes of making and turning cities into smart cities, we can identify several representative features. According to a research study [14], the scientists investigated a period of 10 years of work from 2008 to 2018 and discovered that smart cities might have some features in common. The most interesting features

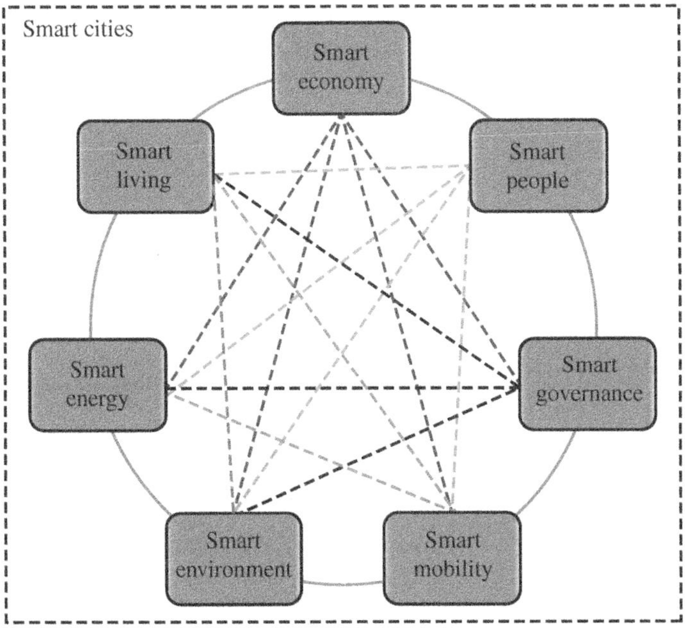

Figure 2.1 Smart cities main features.

in Figure 2.1 [9, 14, 15] are smart economy, smart people, smart governance, smart mobility, smart environment, smart energy, and smart living that are shown in Figure 2.1. As shown in Figure 2.1, although each feature represents the importance of itself to smart cities, they all are interrelated to each other. Each feature has direct or indirect impact on others. Furthermore, proposed decision-making solutions used in smart cities are multi-criteria decision making (MCDM), mathematical programming (MP), AI, and integrated method (IM). In this study, we aim to discover AI solutions, particularly machine learning and deep learning (DL) approaches rather than others, for smart mobility.

2.2 Role of Machine Learning in Smart Cities

Current advanced technologies in sensors and IoT devices make it essential in leveraging AI, particularly machine learning, to model the data for further application [16, 17]. The IoT devices are considered as the most important and unavoidable parts of smart cities. These devices provide a huge amount of data depending on which applications are going to be used, such as healthcare and transportation applications [18].

IoT technologies have proliferated in many fields, such as smart healthcare [13, 19] and smart home systems like Alexa [20, 21], specifically in urban cities, turning them into smart cities. Thus, the huge usage of IoT technologies plays a pivot role in generating big data that requires solutions to analyze and keep track of smart cities' activities. This big data analysis [22] provides invaluable knowledge to integrate all smart cities' sources like IoT and networks. As smart cities and their data grow, this analysis process may become challenging for future decision making. In Section 2.3, we address this problems and discuss possible solutions that have been proposed thus far. Researchers in [23] proposed a new solution to leverage citizen-centered big data analysis to apply to smart cities, which is to determine a path for implementation of citizen-centered big data analysis for the sake of decision making. This solution's main goal is to provide imperative perspectives: data analysis algorithms and urban governance issues [23].

Furthermore, researchers in [24] provided several DL applications in smart cities, such as smart governance, smart urban modeling, smart education, smart transportation, intelligent infrastructures, and smart health solutions. Additionally, the challenges of using DL toward smart cities are also addressed. However, still the problem of decision making in smart cities remains challenging. In Section 2.3, we address the problems and highlight the possible solutions.

2.3 Smart Cities Data Analytics Framework

There are plenty of research studies accomplished in smart cities like [25], which is an intelligent decision computing solution for crowd monitoring. In this section, we dig into smart cities applications, and the foundation of techniques is developed upon by establishing a universal smart cities data analytics framework, which is elaborately depicted in Figure 2.2. This framework has three main sections: data capturing, data analysis, and decision making. We elaborate on each in separate sections.

2.3.1 Data Capturing

Smart cities have been engulfed with too much data that requires the management department to control and monitor the cities, but this department is cost and time inefficient. Hopefully, due to domains (features of smart cities), these data, which are grouped automatically into domains, create particular big data for each domain separately. IoT devices used to gather data for healthcare are completely different from the ones that are developed specially for traffic control (i.e. why we need to categorize the data into groups of domains to create certain technical big data for each domain). According to Figure 2.2, we provide six different samples of sensors

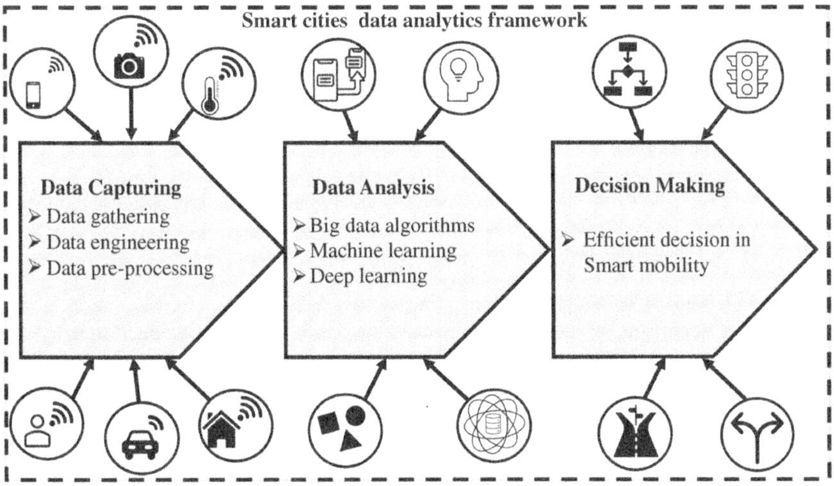

Figure 2.2 Smart cities data analytics framework.

and IoT devices, such as smart phones, smart cameras, smart thermometers, smart users with sensors, smart cars, and smart houses. The variations of sensors enable a data center to receive different types, ranges, and values of data from objects. Here, we highlight the current upcoming challenges with associate solutions.

⋄ **Challenges:** The process of data gathering in smart cities has been enhanced yet remains challenging due to the lack of enough equipment in every place we aim to monitor and make decisions for that region, like relaying traffic information. No matter our focus, whether in smart healthcare mobility or other areas of interest, IoT devices and other sensors may still fail to capture all data correctly or may miss data entirely due to their limited storage and inefficient time-scales.

In addition to that, data have proliferated significantly and are produced from heterogeneous sources. Therefore, the types of data vary from video and images to digits or strings and need particular procedures to convert all of the data into single unit measurements. These measurements enable us to run machine learning algorithms and other DL algorithms on the data readily to make optimum decisions.

⋄ **Solutions:** Preprocessing plays a pivotal role in managing missing information and values within the generated dataset. There are some basic and advanced approaches [26] to handle the missing values, and also, there are tools and techniques to select important and relevant features like IFAB [27] using artificial bee colony to get rid of irrelevant features and ensure the genuineness and reproducibility of the results [28].

To handle the heterogeneous problem, data engineering [22] is responsible for managing and analyzing the input data and adding labels to data left unlabeled.

This requires experts and time; thus it is not cost and time efficient. Therefore, leveraging big data algorithms helps to tune and analyze the data properly.

2.3.2 Data Analysis

The smart cities promises lead us to an ample proliferation and generation in data from all aspects of the domains and branches. Therefore, such huge amounts of data are at the core of the services generated by the IoT technologies [29]. This section of the framework, data analysis, is imperative because its results lead us to make proper decisions. If this process is not accomplished, the decision made will not be efficient. Thus, a large number of research studies have enhanced the process and yielded better results. In the early era of smart cities, there were only limited data generated every day due to the lack of sensors. Therefore, typical machine learning algorithms were sufficient for data analysis to make a model that can handle the situation and provide enough information to make a decision. However, thanks to technologies, the number of sensors and IoT objects have proliferated, and thus we have huge amounts of data that require big data algorithms like and Hadoop to handle the data [30].

Additionally, due to the huge quantity of data, researchers used DL algorithms especially transfer learning and meta-learning [7] and some other famous machine learning techniques to learn within reinforcement learning like Q-learning 31–33 for generating smart systems [34, 35].

2.3.2.1 Big Data Algorithms and Challenges

Due to the big data revolution, the enormous volume of high-performance computations are unavoidable in such smart cities. Thus, big data algorithms are getting one of the important functioning pieces of smart cities.

⋄ **Transportation System:** Transportation systems are generated using advanced technologies where applications of big data vary and are important [36]. Scenarios are established in the research studies to offer the following options to smart people: suggesting the best travel time for any given trip, providing real-time traffic information, predicting movement patterns according to personality (daily routing path) or spatiotemporal routines, enhancing crash analysis, planning bus routes, improving taxi dispatch, and optimizing traffic time during big occasions [36]. For example, Zhu et al. [37] developed an approach to use two algorithms: Bayesian inference and Random forest to execute in real time to predict the probability of crash occurrence to decrease crash risks in smart cities. Moreover, researchers [38] have established a combination of supervised and regression algorithms, such as multivariate adaptive regression splines, regression trees, and logistic regression, to study motor vehicle accident injury behavior.

◇ **Urban Governance**: Learning from human mobility patterns discloses the movement and trend of a large population, the most popular applications of which are in domains like crime prediction [36], disaster evacuation, big events management, and safety estimation. To that end, knowledge about big events derived from bus trajectories, including event start time and end time, can be beneficial for the event planning and management department [39].

2.3.2.2 Machine Learning Process and Challenges

The learning process based on given input where data are not sufficient does not produce reliable and robust results [7]. Machine learning algorithms get stuck in local minimum or maximum easily due to a large amount of data that leads to a problem called overfitting [7]. Furthermore, machine learning algorithms fail to learn positions and states (classes) when the number occurrence in the data is far less than others. For instance, within the smart healthcare domain, when capturing data, the probability of having all desired classes looks low, and the number of rare diseases would be insufficient for learning.

◇ **Solutions:** In a study [40], researchers established an approach called MetaSense, i.e. learn to sense rather than sense to learn. This approach took advantage of one of the important data analytic promises called zero-shot learning (ZSL) [7, 41]. There are many applications of ZSL in computer vision [42] and motion detection in healthcare application [43] Furthermore, researchers in [44] proposed a solution based on ZSL called Zero-virus providing a deep understanding for an intelligent transportation system to generate the best route for drivers. Zero-virus does not need any vehicle-tracklet annotation; thus it is the most volatile real-world traffic scheme.

◇ **Machine Learning Statistic Records:** Iskandaryan et al. [6] investigated a survey to study sensor data and analyze the impact of air quality in smart cities using supervised machine learning algorithms. They established four important and popular categories of algorithms associated with the number of research studies: first, the most popular ML algorithm, neural network (NN); second, logistic regression; third, ensemble algorithm; others. Furthermore, the results in [6] demonstrated that the number of publications associated with machine learning algorithms with the application of smart cities, particularly smart environment by predicting and preventing the air pollution risks, has increased progressively. Additionally, researchers in [6] evaluated the number of metrics that have been used in publications to examine the performance of each algorithm. The two most popular metrics are root mean square error (RMSE) and mean absolute error (MAE).

2.3.2.3 Deep Learning Process and Challenges

Feasible smart cities are established by technology-driven foundations, and their initiatives are on different branches and domains, in which each of them may requires systems with high-performance computing resources and technologies. Such systems have pros and cons such as saving energy and reducing the air pollution and reducing diagnose time, but writing something bad. One of the most popular technologies used to tackle such huge amounts of data is DL that is a special algorithm within machine learning and is efficiently used to obtain required knowledge from the input data, extracting the patterns that govern the whole data and also classify them. Several research studies are successfully applying DL on smart cities [24], such as urban modeling for smart cities, intelligent infrastructure for smart cities, and smart urban governance. Here, we aim to focus on a particular application of DL that is smart mobility and transportation.

AI definitely pushed all the science a step forward by making the systems and processes of scientific inquiry as smart as possible, for example, autonomous transportation systems [24] in smart mobility. Making a decision about whether the object seen is a human being or not is challenging [45]. Object detection is one of the challenging issues in smart mobility that surely boosts and facilitates automation in transportation systems. Consequently, this positively enhances and improves smart mobility in smart cities. Researchers in [45] explored analysis of decent object detection solutions like DL [46]. The scientists leveraged a well-known object detection system, namely, YOLO (You Only Look Once), which was developed earlier by Redmon et al. [47] and assessed its performance on real-time data.

2.3.2.4 Learning Process and Emerging New Type of Data Problems

In this section, we address possible challenges and solutions in the world of data analysis. The first and foremost problem that researchers tackle is lack of data for rare classes within the dataset that are used to make a model. The less number of samples we have in the dataset, the higher chance of ignoring that sample while we train and make the model. To handle this problem, meta-learning has come to play an essential role to make a model only using few samples. It has three important promises: ZSL, one-shot learning, and few-shot learning [7]. ZSL is a certain type of meta-learning when a training dataset does not have any samples for classes, and we still can predict them during a test process. For instance, a research work [44] was conducted to not use any annotation for processing vehicles tracklets. This study established a route understanding system based on zero-shot theory for intelligent transportation, namely, Zero-virus, which obtains high effectiveness with zero samples of annotation of vehicle tracklets. Further, another research work [40] established a new technique, namely MetaSense, which is the process of learning to sense rather than sensing to learn. This process takes advantage of the lack of samples of classes by learning from learning rather than learning from samples.

The process of non-annotation helps to work with data that lacks classes because typical machine learning algorithms fail to detect non-annotation classes. Furthermore, sensing to learn helps new algorithms predict information that is completely lacking from the dataset itself. These advancements lead us closer to (or are part of) ZSL, which is critical for the advancement of Smart Cities.

The second promise is one-shot learning, in which each epoch in a training phase has only one sample per each class that is taken by a DL algorithm or a combination of neural networks [7].

The third promises but not the least one is few (k)-shot learning in which each epoch in a training phase has only few (k) samples per each class that are taken by a DL algorithm or a combination of neural networks [7].

2.3.3 Decision-Making Problems in Smart Cities

Decision-making problems are becoming challenging issue in smart cities where not only the problem itself but also other relevant problems in other aspects of smart cities need to be analyzed. Additionally, decision makers must depict the consequence of the decision they are going to make. Thus, decision-making systems are needed in smart cities in which the systems take care of all issues within the connected networks and only some limited information is taken that would be enough make an optimum decision. In this section, we highlight the challenging decision making problems and solutions.

2.3.3.1 Traffic Decision-Making System

Traffic decision-making system, as known as intelligent transportation system (ITS), aims to detect traffic flow within a smart city and offer optimal solutions using proper big data analytics [48]. Assessing and analyzing this big data plays a pivot role in decision making systems that makes the process time- and cost-efficient [37]. Researchers addressed some case studies within traffic decision-making systems, such as road traffic accidents analysis [49, 50], public transportation management and control, and road traffic flow prediction.

◇ **Challenges:** In these such systems, users feed data into them by gathering data from heterogeneous resources, such as high-performance IoT devices, video detectors, and GPS. The systems must use big data analysis approaches to evaluate the data online [51] to provide efficient and intimate knowledge for decision making. To be more certain about the amount of data these systems handle, consider Petabyte level data that are beyond the utmost of traditional machine learning analytic abilities [51]. This is due to two important issues: these traditional data processing algorithms are not yet developed for online and real-time monitoring systems, and additionally, they fail to learn from these data due to disorganized and nonstandard structures.

◇ **Solutions:** Smart cities, particularly in this domain (i.e. one of their important features, smart mobility, discussed in Figure 2.1), need optimum traffic decision-making systems that address the aforementioned challenges and provide efficient solutions to them. One of the solutions discussed in [51] is the use of big data analytics approaches. They solved the challenges by providing enough data storage, analysis, and management tools. The most important and promising frameworks, libraries, or technologies are used to analyze the big data, such as Hadoop and Spark, but not limited to them. For instance, researchers in [52] established a time-efficient and scalable distributed technique, namely, Partition-Aware Connected Components (PACC), for connected component computation that relies on main approaches: edge filtering, two-step processing of partitioning and computation, and sketching. PACC outperforms MapReduce and Spark frameworks and is time-efficient on real-world graphs [52].

Additionally, big data analysis algorithms facilitate and boost the process of handling large amount of data to provide enough information for making decisions. Traffic decision-making systems using big data algorithms and technologies help associate offices and the people in charge to get to learn drivers' journey patterns within the transportation network that reports whole network trends or better understanding of similar drivers [37]. Considering this feature, the systems provide the best path to drivers to reach their destination through the minimum time possible. Furthermore, the systems help the city to relay the traffic by controlling traffic light, the best timing to decide which light should be on or off and for what period.

Furthermore, traffic decision-making systems predicts the probability of traffic accident occurrence using big data algorithms [37]. This requires to have smart healthcare systems, which is nominated as one of smart cities features discussed in Figure 2.1, to help emergency centers to facilitate the process of emergency rescue.

2.3.3.2 Safe and Smart Environment

Researchers in [5] leveraged DL algorithms and advanced communication technologies to link vehicles, roads, and drivers to facilitate and enhance various traffic-related tasks and improve air pollution [6]. The scientists focused on different initiatives with the goal of creating a safe and smart environment and transportation. Further, Zhu et al. [37] developed an approach to use two algorithms: Bayesian inference and Random forest to execute in real time to predict the probability of crash occurrence to decrease crash risks in smart cities. Moreover, researchers [38] have established a combination of supervised and regression algorithms such as multivariate adaptive regression splines, classification and regression trees, and logistic regression to study the dataset of motor vehicle accident injury.

To have a safe environment, air pollution prediction in smart cities plays an imperative role. There have been a significant amount of work tried to enhance the prediction using several machine learning algorithms. The usage of machine learning algorithms to make environment safe has increased consistently, stating that how important this prediction would be for smart cities [6]. We examined the most relevant research studies [4, 6, 23, 38] but not limited to these, applying different evaluation based on several metrics. The evaluation of those research leads us to the following common observations: first, the rate of applying of the advanced (DL) machine learning algorithms has proliferated rather than typical machine learning algorithms; second, among the prediction elements for air pollution prediction, PM2.5 is considered as the most popular element; third, the data used for air pollution prediction had already generated hourly rather than daily; and in the final observation, efficient prediction occurs when air-quality captured data merged with other relevant data of other networks like healthcare network within the smart cities.

2.4 Conclusion

We highlight smart cities' research branches and technology advancement regarding different complex domains. We propose a solution, namely, universal smart cities' decision making, which has three main sections: data capturing, data analysis, and decision making. We provide an abstract review of the fundamental concepts of big data, ML, and DL algorithms being applied to smart cities. We explore the essential role of the aforementioned algorithms on making decision within smart cities. The goal of this study is to provide a comprehensive survey of data analytics in smart cities, more specifically the role of big data algorithms and other advanced technologies like ML and DL for making decision in smart mobility within smart cities.

References

1 Amini, M.H., Mohammadi, J., and Kar, S. (2020). Promises of fully distributed optimization for IoT-based smart city infrastructures. In: *Optimization, Learning, and Control for Interdependent Complex Networks*, (M Hadi Amini), 15–35. Springer.

2 Montori, F., Bedogni, L., and Bononi, L. (2017). A collaborative internet of things architecture for smart cities and environmental monitoring. *IEEE Internet of Things Journal* 5 (2): 592–605.

3 Bibri, S.E. and Krogstie, J. (2017). Smart sustainable cities of the future: an extensive interdisciplinary literature review. *Sustainable Cities and Society* 31: 183–212.

4 Amini, M.H., Imteaj, A., and Pardalos, P.M. (2020). Interdependent networks: a data science perspective. *Patterns* 1 100003.

5 An, C. and Wu, C. (2020). Traffic big data assisted V2X communications toward smart transportation. *Wireless Networks* 26 (3): 1601–1610.

6 Iskandaryan, D., Ramos, F., and Trilles, S. (2020). Air quality prediction in smart cities using machine learning technologies based on sensor data: a review. *Applied Sciences* 10 (7): 2401.

7 Mohammadi, F.G., Amini, M.H., and Arabnia, H.R. (2020). An introduction to advanced machine learning: meta-learning algorithms, applications, and promises. In: *Optimization, Learning, and Control for Interdependent Complex Networks*, (M Hadi Amini), 129–144. Springer.

8 Neves, F.T., de Castro Neto M., and Aparicio, M. (2020). The impacts of open data initiatives on smart cities: a framework for evaluation and monitoring. *Cities* 106: 102860.

9 Amini, M.H., Arasteh, H., and Siano, P. (2019). Sustainable smart cities through the lens of complex interdependent infrastructures: panorama and state-of-the-art. In: *Sustainable Interdependent Networks II*, (M Hadi Amini), 45–68. Springer.

10 Park, J.H., Younas, M., Arabnia, H.R., and Chilamkurti, N. (2021). Emerging ICT applications and services—big data, IoT, and cloud computing. *International journal of communication systems*, 34 e4668.

11 Hossain, M.S., Muhammad, G., and Alamri, A. (2019). Smart healthcare monitoring: a voice pathology detection paradigm for smart cities. *Multimedia Systems* 25 (5): 565–575.

12 Rocha, N.P., Dias, A., Santinha, G. et al. (2019). Smart cities and healthcare: a systematic review. *Technologies* 7 (3): 58.

13 Ellaji, Ch., Sreehitha, G., and Devi, B.L. (2020). Efficient health care systems using intelligent things using NB-IoT. *Materials Today: Proceedings*.

14 Hoang, G.T.T., Dupont, L., and Camargo, M. (2019). Application of decision-making methods in smart city projects: a systematic literature review. *Smart Cities* 2 (3): 433–452.

15 de Oliveira, L.F.P., Manera, L.T., and Luz, P.D.G. (2020). Development of a smart traffic light control system with real-time monitoring. *IEEE Internet of Things Journal.* 8 3384–3393.

16 Boulos, M.N.K., Peng, G., and VoPham, T. (2019). An overview of GeoAI applications in health and healthcare. *International Journal of Health Geographics*, 18 1–9.

17 Al-Turjman, F., Nawaz, M.H., and Ulusar, U.D. (2020). Intelligence in the internet of medical things era: a systematic review of current and future trends. *Computer Communications* 150: 644–660.

18 Ullah, Z., Al-Turjman, F., Mostarda, L., and Gagliardi, R. Applications of artificial intelligence and machine learning in smart cities. (2020). *Computer Communications*. 154 313–323.

19 Manikandan, R., Patan, R., Gandomi, A.H. et al. (2020). Hash polynomial two factor decision tree using IoT for smart health care scheduling. *Expert Systems with Applications* 141: 112924.

20 Khatri, C., Hedayatnia, B., Venkatesh, A. et al. (2018). Advancing the state of the art in open domain dialog systems through the Alexa prize. *arXiv preprint arXiv:1812.10757*.

21 Rak, M., Salzillo, G., and Romeo, C. (2020). Systematic IoT penetration testing: Alexa case study. *ITASEC*, pp. 190–200.

22 Elhoseny, H., Elhoseny, M., Riad, A.M., and Hassanien, A.E. (2018). A framework for big data analysis in smart cities. *International Conference on Advanced Machine Learning Technologies and Applications*, Springer, pp. 405–414.

23 Ju, J., Liu, L., and Feng, Y. (2018). Citizen-centered big data analysis-driven governance intelligence framework for smart cities. *Telecommunications Policy* 42 (10): 881–896.

24 Bhattacharya, S., Somayaji, S.R.K., Gadekallu, T.R. et al. (2020). A review on deep learning for future smart cities. *Internet Technology Letters* e187.

25 Kumar, S., Datta, D., Singh, S.K., and Sangaiah, A.K. (2018). An intelligent decision computing paradigm for crowd monitoring in the smart city. *Journal of Parallel and Distributed Computing* 118: 344–358.

26 Usman, M., Jan, M.A., He, X., and Chen, J. (2019). A survey on big multimedia data processing and management in smart cities. *ACM Computing Surveys (CSUR)* 52 (3): 1–29.

27 Mohammadi, F.G. and Abadeh, M.S. (2014). Image steganalysis using a bee colony based feature selection algorithm. *Engineering Applications of Artificial Intelligence* 31: 35–43.

28 Shenavarmasouleh, F. and Arabnia, H. (2019). Causes of misleading statistics and research results irreproducibility: a concise review. *2019 International Conference on Computational Science and Computational Intelligence (CSCI)*, pp. 465–470.

29 Hashem, I.A.T., Chang, V., Anuar, N.B. et al. (2016). The role of big data in smart city. *International Journal of Information Management* 36 (5): 748–758.

30 Mohammadi, F.G., Shenavarmasouleh, F., Arabnia, H.R., and Amini, M.H. (2020). Impact of weather conditions on the Covid-19 pandemic in the United States: a big data approach. *2020 International Conference on Computational Science and Computational Intelligence (CSCI)*. IEEE.

31 Chakole, J.B., Kolhe, M.S., Mahapurush, G.D. et al. (2021). A Q-learning agent for automated trading in equity stock markets. *Expert Systems with Applications* 163: 113761.

32 Liao, Z., Peng, J., Chen, Y. et al. (2020). A fast Q-learning based data storage optimization for low latency in data center networks. *IEEE Access* 8: 90630–90639.

33 Fan, J., Wang, Z., Xie, Y., and Yang, Z. (2020). A theoretical analysis of deep Q-learning. *Learning for Dynamics and Control*, PMLR, pp. 486–489.

34 Boussakssou, M., Hssina, B., and Erittali, M. (2020). Towards an adaptive E-learning system based on Q-learning algorithm. *Procedia Computer Science* 170: 1198–1203.

35 Joo, H., Ahmed, S.H., and Lim, Y. (2020). Traffic signal control for smart cities using reinforcement learning. *Computer Communications* 154: 324–330.

36 Wang, A., Zhang, A., Chan, E.H.W. et al. (2021). A review of human mobility research based on big data and its implication for smart city development. *ISPRS International Journal of Geo-Information* 10 (1): 13.

37 Zhu, L., Yu, F.R., Wang, Y. et al. (2018). Big data analytics in intelligent transportation systems: a survey. *IEEE Transactions on Intelligent Transportation Systems* 20 (1): 383–398.

38 Xiong, G., Zhu, F., Fan, H. et al. (2014). Novel its based on space-air-ground collected big-data. *17th International IEEE Conference on Intelligent Transportation Systems (ITSC)*, IEEE, pp. 1509–1514.

39 Mazimpaka, J.D. and Timpf, S. (2017). How they move reveals what is happening: understanding the dynamics of big events from human mobility pattern. *ISPRS International Journal of Geo-Information* 6 (1): 15.

40 Mohammadi, F.G. and Amini, M.H. (2019). Promises of meta-learning for device-free human sensing: learn to sense. *Proceedings of the 1st ACM International Workshop on Device-Free Human Sensing*, pp. 44–47.

41 Wei, K., Deng, C., and Yang, X. (2020). Lifelong zero-shot learning. *Proceedings of the 29th International Joint Conference on Artificial Intelligence, IJCAI-20*, pp. 551–557.

42 Mohammadi, F.G., Arabnia, H.R., and Amini, M.H. (2019). On parameter tuning in meta-learning for computer vision. *2019 International Conference on Computational Science and Computational Intelligence (CSCI)*, IEEE, pp. 300–305.

43 Wijekoon, A., Wiratunga, N., and Sani, S. (2018). Zero-shot learning with matching networks for open-ended human activity recognition. *CEUR Workshop Proceedings.*

44 Yu, L., Feng, Q., Qian, Y. et al. (2020). Zero-virus: zero-shot vehicle route understanding system for intelligent transportation. *Proceedings of the IEEE/CVF Conference on Computer Vision and Pattern Recognition Workshops*, pp. 594–595.

45 Pandey, A., Puri, M., and Varde, A. (2018). Object detection with neural models, deep learning and common sense to aid smart mobility. *2018 IEEE 30th International Conference on Tools with Artificial Intelligence (ICTAI)*, IEEE, pp. 859–863.

46 Asali, E., Shenavarmasouleh, F., Mohammadi, F.G. et al. (2020). DeepMSRF: A novel deep multimodal speaker recognition framework with feature selection. *ArXiv*, abs/2007.06809.

47 Redmon, J., Divvala, S., Girshick, R., and Farhadi, A. (2016). You only look once: unified, real-time object detection. *Proceedings of the IEEE Conference on Computer Vision and Pattern Recognition*, pp. 779–788.

48 Safari, Z., Mursi, K.T., and Zhuang, Y. (2020). Fast automatic determination of cluster numbers for high dimensional big data. *Proceedings of the 2020 the 4th International Conference on Compute and Data Analysis*, pp. 50–57.

49 Golob, T.F. and Recker, W.W. (2003). Relationships among urban freeway accidents, traffic flow, weather, and lighting conditions. *Journal of Transportation Engineering* 129 (4): 342–353.

50 Xu, J., Deng, D., Demiryurek, U. et al. (2015). Mining the situation: spatiotemporal traffic prediction with big data. *IEEE Journal of Selected Topics in Signal Processing* 9 (4): 702–715.

51 Mohamed, N. and Al-Jaroodi, J. (2014). Real-time big data analytics: applications and challenges. *2014 International Conference on High Performance Computing & Simulation (HPCS)*, IEEE, pp. 305–310.

52 Park, H.-M., Park, N., Myaeng, S.-H., and Kang, U. (2020). PACC: Large scale connected component computation on Hadoop and Spark. *PLoS ONE* 15 (3): e0229936.

3

Embodied AI-Driven Operation of Smart Cities: A Concise Review

Farzan Shenavarmasouleh[1], *Ghareh Mohammadi*[1], *M. Hadi Amini*[2], *and Hamid Reza Arabnia*[1]

[1] *Department of Computer Science, Franklin College of Arts and Sciences, University of Georgia, Athens, GA, USA*
[2] *Knight Foundation School of Computing and Information Sciences, Florida International University, Miami, FL, USA*

3.1 Introduction

A smart city is an urban area that employs information and communication technologies (ICT) [1], an intelligent network of connected devices and sensors that can work interdependently [2, 3] and a distributive manner [4] to continuously monitor the environment, collect data, and share them among the other assets in the ecosystem. This uses all the available data to make real-time decisions about the many individual components of the city to ease up the livelihood of its citizens and make the whole system more efficient, more environmentally friendly, and more sustainable [5]. This serves as a catalyst for creating a city with faster transportation, fewer accidents, enhanced manufacturing, more reliable medical services and utilities, less pollution [6], and so on. The good news is any city, even with traditional infrastructures, can be transformed into a smart city by integrating Internet of things (IoT) technologies [7].

An undeniable part of a smart city is its use of smart agents. These agents can vary a lot in sizes, shapes, and functionalities. They can simply be light sensors that along with their controller act as the energy-saving agents or could be more advanced machines, with complicated controllers and interconnected components that are capable of tackling more advanced problems. The latter agents usually come with an embodiment with numerous sensors and controllers built in them, enabling them to perform high-level and human-level tasks such as talking, walking, seeing, and complex reasoning along with the ability to interact with the environment. Embodied artificial intelligence is the field of study that

Cyberphysical Smart Cities Infrastructures: Optimal Operation and Intelligent Decision Making, First Edition. Edited by M. Hadi Amini and Miadreza Shafie-khah.

takes a deeper look into these agents and explores how they can fit into the real-world and how they can eventually act as our future community workers, personal assistants, robocops, and many more.

Imagine arriving home after a long working day and seeing your home robot waiting for you at the entrance door. Although it is not the most romantic thing ever, you then walk up to it, and ask it to make a cup of coffee for you and also add two teaspoons of sugar if there is any in the cabinet. For this to become reality, the robot has to have a vast range of skills. It should be able to understand your language and be able to translate questions and instructions to the action. It should be able to see its surroundings and have the ability to recognize objects and scenes. Last but not the least, it must know how to navigate in a big dynamic environment, interact with the objects within it, and be capable of doing long-term planning and reasoning.

In the past few years, there has been significant progress in the fields of computer vision, natural language processing, and reinforcement learning, thanks to the advancements in deep learning models. Many things are now possible because of these that seemed impossible a few years ago. However, most of the work has been done in isolation from other lines of work. It means that the trained model can only take one type of data (e.g. image, text, video) as the input and perform a single task that it is asked for. Consequently, such a model acts as a single-sensory machine as opposed to a multi-sensory one. Also, for the most part, they all belong to Internet artificial intelligence (AI) rather than embodied AI. The goal of Internet AI is learning patterns in text, images, and videos from the datasets collected from the Internet.

If we zoom out and look at the way models in Internet AI being trained, we realize that generally supervised classification is the way to go. For instance, we provide a certain number of dog and cat photos along with the corresponding labels to a perception model. Moreover, if the number is large enough, the model then can successfully learn the differences between these two animals and discriminate between them. Learning via flashcards falls under the same umbrella for humans.

Extensive amount of time has been devoted in the past years to gather and build huge datasets for the imaging and language communities. A few considerable markers of this can be ImageNet [8], MS COCO [9], Sun [10], Caltech-256 [11], and Places [12] created for vision tasks; SQuAD [13], GLUE [14], and SWAG [15] built for language objectives; and Visual Genome [16] and VQA [17] datasets created for joint purposes to name a few.

Apart from playing a pivotal role in the recent advances of the main fields, these datasets also proved to be useful when used with transfer learning methods to help underlying disciplines such as biomedical imaging [18, 19]. However, the aforementioned datasets are prune to restrictions. Firstly, at times it can get extremely costly, both in terms of time and money, to gather all the required data for the

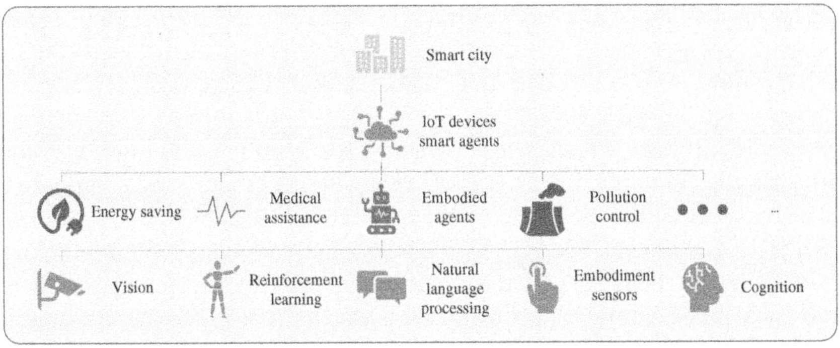

Figure 3.1 Embodied AI in smart cities.

collection and label them. Secondly, the collection has to be monitored constantly to assure that they follow certain rules to avoid creating biases that could lead to erroneous results in future works [20] and also make sure that the collected data are all normal and uniform in terms of attributes such as background, size, position of the objects, lighting conditions, etc. However, in contrast, we know that in real-world scenarios, this cannot be the case, and robots have to deal with a mixture of unnormalized noisy irrelevant data along with the relevant well-curated ones. Additionally, the agent would be able to interact with the objects in the wild (e.g. picking it up and looking at the object from another angle) and also use its other senses such as smell and hearing to collect information (Figure 3.1).

Humans do learn from interactions, and it is a must for true intelligence in the real world. In fact, it is not only humans but also animals. In kitten carousel experiment [21], Held and Hein exhibited this beautifully. They studied the visual development of two kittens in a carousel over time. One of them had the ability to touch the ground and control its motions within the restrictions of the device, while the other was just a passive observer. At the end of the experiment, they found out that the visual development of the former kitten was normal, whereas for the latter one it was not, even though they both saw the same thing. This proves that being able to physically experience the world and interact with it is a key element for learning [22].

The goal of embodied AI is to bring the ability to interact and being able to use multisenses simultaneously into play to enable the robot to continuously learn in a lightly supervised or even unsupervised way in a rich dynamic environment.

3.2 Rise of the Embodied AI

In the mid-1980s, a major paradigm shift took place toward embodiment, and computer science started to become more practical than theoretical algorithms and approaches. Embedded systems started to appear in all kinds of forms to aid humans in everyday life. Controllers for trains, airplanes, elevators, air conditioners, and software for translation and audio manipulation are some of the most important ones, to name a few [23].

Embodied artificial intelligence is a broad term, and those successes were for sure great ones to start with. Yet, it could clearly be seen that it was a huge room for improvement. Theoretically, the ultimate goal of AI is to not only master any given algorithm or task that is given to but also gain the ability to multitask and get to human-level intelligence, and that as mentioned requires meaningful interaction with the real world. There are many specialized robots for a vast set of tasks out there, especially in large industries, which can do the assigned task to perfection, let it be cutting different metals, painting, soldering circuits, and many more. However, until one single machine emerges to have the ability to do different tasks or at least a small subset of them by itself and not just by following orders, it cannot be called intelligence.

Humanoids are the main thing that comes to mind when we talk about robots with intelligence. Although it is the ultimate goal, it is not the only form of intelligence on Earth. Other animals, such as insects, have their own kind of intelligence, and due to being relatively simpler compared to humans, they are a very good place to begin with.

Rodney Brooks has a famous argument that says it took the evolution much longer to create insects from scratch than getting to human-level intelligence from there. Consequently, he suggested that these simpler biorobotics should be first dealt with in the road to make much more complex ones. Genghis, a six-legged walking robot [24], is one of his contributions to this field.

This line of thought was a fundamental change and led researchers to have a change of direction in their work, and with that came attention to new domains and topics such as robotics, locomotion, artificial life, bioinspired systems, and so on. The classical approach did not care about tasks related to the interaction with the real world, and consequently, this journey is started by locomotion and grasping.

Since not much computational power was available at the time of this shift, a big challenge for the researchers was the trade-off between simplicity and the potential to operate in complex environments. An extensive amount of work has been done in this area to explore or invent ways to exploit natural body dynamics, materials used in the modules, and their morphologies to make the robots move and become able to grasp and manipulate items without sophisticated processing units [25, 27]. It goes without saying that the ones who could use the physical properties of themselves and the environment to function were more energy-efficient, but they had their own limitations. Not being able to generalize well to complex environments was a major drawback. However, they were fast as the machines with huge processing units needed a reasonable amount of time to think and plan their next action and often move their rigid and non-smooth actuators.

Nowadays, a big part of these issues are solved, and we can see extremely fast and smooth natural moving robots capable of doing different types of maneuvers [28], but yet it is foreseen that with the advances of artificial muscles, joints, and tendons, this progress can be further improved.

3.3 Breakdown of Embodied AI

In this section, we try to categorize a broad range of research that has been done under the field of embodied AI. Due to the huge diversity, each section will necessarily be abstract and selective and reflect the authors' personal opinion.

3.3.1 Language Grounding

Machine and human communication has always been a topic of interest. As time goes on, more and more aspects of our lives are controlled by AIs, and hence it is crucial to have ways to talk with them. This is a must for giving new instructions to them or receiving an answer from them, and since we are talking about general day-to-day machines, we desire this interface to be higher level than programming languages and closer to spoken language. To achieve this, machines must be capable of relating language to actions and the world. Language grounding is the field that tries to tackle this and map natural language instructions to robot behavior.

Hermann et al.'s study shows that this can be achieved by rewarding an agent upon successful execution of written instructions in a 3D environment with a combination of unsupervised learning and reinforcement learning [29]. They also argue that their agent can generalize well after training and can interpret new unseen instructions and operate in unfamiliar situations.

3.3.2 Language Plus Vision

Now that we know that machines can understand languages and there exist sophisticated models just for this purpose out there [30], it is time to bring another sense into play. One of the most popular ways to show the potential of joint training of vision and language is the image and video captioning [31, 35].

More recently, a new line of work has been introduced to take advantage of this connection. AbbtextVisual question answering (VQA) [17] is the task of receiving an image along with a natural language question about that image as an input and attempting to find the accurate natural language answer for it as the output. The beauty of this task is that both the questions and the answers can be open-ended and also the questions can target different aspects of the image such as the objects that are present in them, their relationship or relative positions, colors, and background.

Following this research, Singh et al. [36] cleverly added an optical character recognition (OCR) module to the VQA model to enable the agent to read the texts available in the image as well and answer questions asked from them or use the additional context indirectly to answer the question better.

One may ask where the new task stands relative to the previous one. Do agents who can answer questions more intelligent than the ones who deal with captions or not? The answer is yes. In [17], the authors show that VQA agents need a deeper and more detailed understanding of the image and reasoning than models for captioning.

3.3.3 Embodied Visual Recognition

Passive or fixed agents may fail to recognize objects in scenes if they are partially or heavily occluded. Embodiment comes to the rescue here and gifts the possibility of moving in the environment to actively control the viewing position and angle to remove any ambiguity in object shapes and semantics.

Jayaraman and Grauman [37] started to learn representations that will exploit the link between how the agent moves and how it will affect its visual surrounding. To do this they used raw unlabeled videos along with an external GPS sensor that provided the agent's coordinates and trained their model to learn a representation linking these two. So, after this, the agent would have the ability to predict the outcome of its future actions and guess how the scene would look like after moving forward or turning to a side.

This was powerful and in a sense, the agent developed imagination. However, there was an issue here. If we pay attention, we realize that the agent is still being fed prerecorded video as the input and is learning similar to the observer kitten in the kitten carousel experiment explained above. So, following this, the authors went after this problem and proposed to train an agent that takes any given object from an arbitrary angle and then predict or better to say imagine the other views by finding the representation in a self-supervised manner [38].

Up until this point, the agent does not use the sound of its surroundings while humans are all about experiencing the world in a multisensory manner. We can see, hear, smell, and touch all at the same time and extract and use the relevant information that could be beneficial to our task at hand. All that said, understanding and learning the sound of objects present in a scene is not easy since all the sounds are overlapped and are being received via a single channel sensor. This is often dealt with as an audio source separation problem, and lots of work has been done on it in the literature [39, 43].

Now it was the reinforcement learning's turn to make a difference. Policies have to be learned to help agents move around a scene, and this is the task of active recognition [44, 48]). The policy will be learned at the same time it is learning other tasks and representation, and it will tell the agent where and how to strategically move to recognize things faster [49, 50].

Results show that policies indeed help the agent to achieve better visual recognition performance, and the agents can strategize their future moves and path for better results that are mostly different from shortest paths [51].

3.3.4 Embodied Question Answering

Embodied Question Answering brings QA into the embodied world. The task starts by an agent being spawned at a random location in a 3D environment and asked a question in which its answer can be found somewhere in the environment. For the agent to answer it, it must first strategically navigate to explore the environment, gather necessary data via its vision, and then answer the question when the agent finds it [52, 53].

Following this, Das et al. [54] also presented a modular approach to further enhance this process by teaching the agent to break the master policy into subgoals that are also interpretable by humans and execute them to answer the question. This proved to increase the success rate.

3.3.5 Interactive Question Answering

Interactive Question Answering (IQA) is closely related to the Embodied version of it. The only main issue is that question is designed in a way that the agent must interact with the environment to find the answer. For example, it has to open the refrigerator or pick up something from the cabinet and then plan for a series of actions conditioned on the question [55].

3.3.6 Multi-agent Systems

Multi-agent systems (MAS) is another interesting line of development. The default standpoint of AI has a strong focus on individual agents. MAS research that has its origins in the field of biology tries to change this and studies the emergence of behaviors in groups of agents or swarms instead [56, 57].

Every agent has a set of abilities and is good in them to an extent. The point of interest in MAS is how a sophisticated global behavior can emerge from a population of agents working together. A real-life example of such behavior can be found in insects like ants and bees [58, 59]. One of the interesting goals of this research is to ultimately make agents that could self-repair [60, 61].

The emerging behavior of MAS can be tailored by researchers to let the group of agents tackle various tasks such as rescue missions, traffic control, fun sports events, surveillance, and many more. Additionally, when fused with other fields, unexpected outcomes can occur. Take "Talking Heads" experiment by Luc Steels [62, 63] as an example showing a common vocabulary emerging through the interaction of agents with each other and their environment via a language game.

3.4 Simulators

Now that we know about the fields and tasks that embodied AI can shine in, the question is how our agents should be trained. One may say it is good to directly train in the physical world and expose them to its richness. Although a valid solution, this choice comes with a few drawbacks. First, the training process in the real world is slow, and the process cannot be sped up or parallelized. Second, it is very hard to control the environment and create custom scenarios. Third, it is expensive, both in terms of power and time. Fourth, it is not safe, and improperly trained or not fully trained robots can hurt themselves, humans, animals, and other assets. Fifth, for the agent to generalize the training, it has to be done in plenty of different environments that is not feasible in this case.

Our next choice is simulators, which can successfully deal with all the aforementioned problems pretty well. In the shift from Internet AI to embodied AI, simulators take the role that was previously played by traditional datasets. Additionally, one more advantage of using simulators is that the physics in the environment can be tweaked as well. For instance, some traditional approaches in this field [64] are sensitive to noise, and for the remedy, the noise in the sensors can be turned off for the purpose of this task.

As a result, agents nowadays are often developed and benchmarked in simulators [65, 66], and once a promising model has been trained and tested, it can then be transferred to the physical world [67, 68].

House3D [69], AI2-THOR [70], Gibson [71], CHALET [72], MINOS [73], and Habitat [74] are some of the popular simulators for the embodied AI studies. These platforms vary with respect to the 3D environments they use, the tasks they can handle, and the evaluation protocols they provide. These simulators support different sensors such as vision, depth, touch, and semantic segmentation.

In this chapter, we mainly focus on MINOS and Habitat since they provide more customization abilities (number of sensors, their positions, and their parameters) and are implemented in a loosely coupled manner to generalize well to new multisensory tasks and environments. As their API can be used to define any high-level task and the material, object clutter variation and many more can be programmatically configured for the environment. They both support navigation with both continuous and discrete state spaces. Also, for the purpose of their benchmarks, all the actuators are noiseless, but they both have the ability to enable noises if desired [75].

In the last section, we saw numerous task definitions and how they each can be tackled by the agents. So, before jumping into MINOS and Habitat simulators and reviewing them, let us first get more familiarized with the three main goal-directed navigation tasks, namely, PointGoal Navigation, ObjectGoal Navigation, and RoomGoal Navigation.

In PointGoal Navigation, an agent is appeared at a random starting position and orientation in a 3D environment and is asked to navigate to target coordinates that are given relative to the agent's position. The agent can access its position via an indoor GPS. There exists no ground-truth map, and the agent must only use its sensors to do the task. The scenarios start the same for ObjectGoal Navigation and RoomGoal Navigation as well; however, instead of coordinates, the agent is asked to find an object or go to a specific room.

3.4.1 MINOS

MINOS simulator provides access to 45 000 three-dimensional models of furnished houses with more than 750 K rooms of different types available in the SUNCG [76] dataset and 90 multi-floor residences with approximately 2000 annotated room regions that are in the Matterport3D [77] dataset by default. Environments in Matterport3D are more realistic looking than the ones in SUNCG. MINOS simulator can approximately reach hundreds of frames per second on a normal workstation.

To benchmark the system, the authors studied four navigation algorithms, three of which were based on asynchronous advantage actor-critic (A3C) approach [78] and the remaining one was direct future prediction (DFP) [79].

The most basic one among the algorithms was feedforward A3C. In this algorithm, a feedforward convolutional neural network (CNN) model is employed as the function approximator to learn the policy along with the total value function that is the expected sum of rewards from the current timestamp until the end of the episode. The second one was LSTM A3C that used an LSTM model with the feedforward A3C acting as a simple memory. Next was UNREAL, an LSTM A3C model boosted with auxiliary tasks such as value function replay and reward prediction. Last but not the least, the DFP algorithm was employed that can be considered as Monte Carlo RL [80] with a decomposed reward.

The authors benchmarked these algorithms on PointGoal and RoomGoal tasks and found out that, firstly, the naive feedforward algorithm fails to learn any useful representation and, secondly, in small environments, DFP performs better while in big and more complex environments, UNREAL beat the others.

3.4.2 Habitat

Habitat was designed and built in a way to provide the maximum customizability in terms of the datasets that can be used and how the agents and the environment can be configured. That being said, Habitat works with all the major 3D environment datasets without a problem. Moreover, it is extremely fast in comparison with other simulators. AI2-THOR and CHALET can get to an fps of roughly 10, MINOS and Gibson can get to around a hundred, and House3D yields 300 fps in the best case, while Habitat is capable of getting up to 10 000 fps. It also provides a more realistic collision model in which if a collision happens, the agent can be moved partially or not at all in the intended direction.

To benchmark Habitat, the owners employed a few naive algorithm baselines, proximal policy optimization (PPO) [81] as the representer of learning algorithms versus ORB-SLAM2 [82, 83] as the chosen candidate for non-learning agents, and tested them on the PointGoal Navigation task on Gibson and Matterport3D. They used Success weighted by Path Length (SPL) [84] as the metric for their performance. The PPO agent was tested with different levels of sensors (e.g. no visual sensor, only depth, only RGB, and RGBD) to perform an ablation study and find the proportion in which each sensor helps the progress. SLAM agents were given RGBD sensors in all the episodes.

The authors found out that first, PPO agents with only RGB perform as bad as agents with no visual sensors. Second, all agents perform better and generalize more on Gibson rather than Matterport3D since the size of environments in the latter is bigger. Third, agents with only depth sensors generalize across datasets the best and can achieve the highest SPL. However, most importantly, they realized that unlike what has been mentioned in the previous work, if the PPO agent learns long enough, it will eventually outperform the traditional SLAM pipeline. This finding was only possible because the Habitat simulator was fast enough to train PPO agents for 75 million time steps as opposed to only 5 million time steps in the previous investigations.

3.5 Future of Embodied AI

3.5.1 Higher Intelligence

Consciousness has always been considered as the ultimate characteristic for true intelligence. Qualia [85, 86] is the philosophical view of consciousness and is related to the subjective sensory qualities like "the redness of red" that humans have in their mind. If at some point machines can understand this concept and objectively measure such things, then the ultimate goal can be marked as accomplished.

Robots still struggle at performing a wide spectrum of tasks effortlessly and smoothly, and this is mainly due to actuator technology as currently most electrical motors are used. Advances in artificial muscles and skin sensors that could cover the entire embodiment of the agent would be essential to fully mitigate the human experience in the real world and eventually unlock the desired cognition [87].

3.5.2 Evolution

One more key component for cognition is the ability to grow and evolve over time [88, 90]. It is easy to evolve the agent's controller via an evolutionary algorithm, but it is not enough. If we aim to have completely different agents, we might as well give them the ability to evolve in terms of embodiment and the sensors as well. This again requires the abovementioned artificial cell organism to encode different physical attributes in them and flip them slightly over time. Of course, we are far from this to become reality, but it is always good to know the furthermost step that has to be done one day.

3.6 Conclusion

Embodied AI is the field of study that takes us one step closer to the true intelligence. It is a shift from Internet AI toward embodiment intelligence that tries to exploit the multisensory abilities of agents such as vision, hearing, and touch, together with language understanding and reinforcement learning attempts to interact in the real world in a more sensible way. In this chapter, we tried to do a concise review of this field and its current advancements, subfields, and tools expecting that this would help accelerate future researches in this area.

References

1 Park, J.H., Younas, M., Arabnia, H.R., and Chilamkurti, N. (2021). Emerging ICT applications and services-big data, IoT, and cloud computing. *International Journal of Communication Systems*. https://onlinelibrary.wiley.com/doi/full/10.1002/dac.4668.

2 Amini, M.H., Imteaj, A., and Pardalos, P.M. (2020). Interdependent networks: a data science perspective. *Patterns* 1 100003. https://www.sciencedirect.com/science/article/pii/S2666389920300039.

3 Mohammadi, F.G. and Amini, M.H. (2019). Promises of meta-learning for device-free human sensing: learn to sense. *Proceedings of the 1st ACM International Workshop on Device-Free Human Sensing*, pp. 44–47.

4 Amini, M.H., Mohammadi, J., and Kar, S. (2020). Promises of fully distributed optimization for IoT-based smart city infrastructures. In: *Optimization, Learning, and Control for Interdependent Complex Networks*, M. Hadi Amini, 15–35. Springer.

5 Amini, M.H., Arasteh, H., and Siano, P. (2019). Sustainable smart cities through the lens of complex interdependent infrastructures: panorama and state-of-the-art. In: *Sustainable Interdependent Networks II*, (M. Hadi Amini, Kianoosh G. Boroojeni, S. S. Iyengar et al.), 45–68. Springer.

6 Iskandaryan, D., Ramos, F., and Trilles, S. (2020). Air quality prediction in smart cities using machine learning technologies based on sensor data: a review. *Applied Sciences* 10 (7): 2401.

7 Batty, M., Axhausen, K.W., Giannotti, F. et al. (2012). Smart cities of the future. *The European Physical Journal Special Topics* 214 (1): 481–518.

8 Deng, J., Dong, W., Socher, R. et al. (2009). ImageNet: A large-scale hierarchical image database. *2009 IEEE Conference on Computer Vision and Pattern Recognition*, IEEE, pp. 248–255.

9 Lin, T.-Y., Maire, M., Belongie, S. et al. (2014). Microsoft COCO: Common objects in context. *European Conference on Computer Vision*, Springer, pp. 740–755.

10 Xiao, J., Hays, J., Ehinger, K.A. et al. (2010). Sun database: large-scale scene recognition from abbey to zoo. *2010 IEEE Computer Society Conference on Computer Vision and Pattern Recognition*, IEEE, pp. 3485–3492.

11 Griffin, G., Holub, A., and Perona, P. (2007). Caltech-256 object category dataset.

12 Zhou, B., Lapedriza, A., Xiao, J. et al. (2014). Learning deep features for scene recognition using places database. *Advances in Neural Information Processing Systems* 27: 487–495.

13 Rajpurkar, P., Zhang, J., Lopyrev, K., and Liang, P. (2016). SQuAD: 100,000+ questions for machine comprehension of text. *arXiv preprint arXiv:1606.05250.*

14 Wang, A., Singh, A., Michael, J. et al. (2018). GLUE: A multi-task benchmark and analysis platform for natural language understanding. *arXiv preprint arXiv:1804.07461.*

15 Zellers, R., Bisk, Y., Schwartz, R., and Choi, Y. (2018). SWAG: A large-scale adversarial dataset for grounded commonsense inference. *arXiv preprint arXiv:1808.05326.*

16 Krishna, R., Zhu, Y., Groth, O. et al. (2017). Visual genome: connecting language and vision using crowdsourced dense image annotations. *International Journal of Computer Vision* 123 (1): 32–73.

17 Antol, S., Agrawal, A., Lu, J. et al. (2015). VQA: Visual question answering. *Proceedings of the IEEE International Conference on Computer Vision*, pp. 2425–2433.

18 Shenavarmasouleh, F. and Arabnia, H.R. (2020). DRDr: Automatic masking of exudates and microaneurysms caused by diabetic retinopathy using mask R-CNN and transfer learning. *arXiv preprint arXiv:2007.02026.*

19 Shenavarmasouleh, F., Mohammadi, F.G., Amini, M.H., and Arabnia, H.R. (2020). DRDr II: Detecting the severity level of diabetic retinopathy using mask RCNN and transfer learning. *arXiv preprint arXiv:2011.14733.*

20 Shenavarmasouleh, F. and Arabnia, H. (2019). Causes of misleading statistics and research results irreproducibility: a concise review. *2019 International Conference on Computational Science and Computational Intelligence (CSCI)*, pp. 465–470.

21 Held, R. and Hein, A. (1963). Movement-produced stimulation in the development of visually guided behavior. *Journal of Comparative and Physiological Psychology* 56 (5): 872.

22 Moravec, H. (1984). Locomotion, vision and intelligence.

23 Hoffmann, M. and Pfeifer, R. (2012). The implications of embodiment for behavior and cognition: animal and robotic case studies. *arXiv preprint arXiv:1202.0440.*

24 Brooks, R.A. (1991). New approaches to robotics. *Science* 253 (5025): 1227–1232.

25 Collins, S.H., Wisse, M., and Ruina, A. (2001). A three-dimensional passive-dynamic walking robot with two legs and knees. *The International Journal of Robotics Research* 20 (7): 607–615.

26 Iida, F. and Pfeifer, R. (2004). Cheap rapid locomotion of a quadruped robot: self-stabilization of bounding gait. In: *Intelligent Autonomous Systems*, vol. 8, 642–649. The Netherlands: IOS Press Amsterdam.

27 Yamamoto, T. and Kuniyoshi, Y. (2001). Harnessing the robot's body dynamics: a global dynamics approach. *Proceedings 2001 IEEE/RSJ International Conference on Intelligent Robots and Systems. Expanding the Societal Role of Robotics in the Next Millennium (Cat. No. 01CH37180)*, Volume 1, IEEE, pp. 518–525.

28 Bledt, G., Powell, M.J., Katz, B. et al. (2018). MIT cheetah 3: design and control of a robust, dynamic quadruped robot. *2018 IEEE/RSJ International Conference on Intelligent Robots and Systems (IROS)*, IEEE, pp. 2245–2252.

29 Hermann, K.M., Hill, F., Green, S. et al. (2017). Grounded language learning in a simulated 3D world. *arXiv preprint arXiv:1706.06551.*

30 Tenney, I., Das, D., and Pavlick, E. (2019). Bert rediscovers the classical NLP pipeline.

31 Pan, Y., Yao, T., Li, H., and Mei, T. (2017). Video captioning with transferred semantic attributes. *Proceedings of the IEEE Conference on Computer Vision and Pattern Recognition*, pp. 6504–6512.

32 Amirian, S., Rasheed, K., Taha, T.R., and Arabnia, H.R. (2020). Automatic image and video caption generation with deep learning: a concise review and algorithmic overlap. *IEEE Access* 8: 218386–218400.

33 Amirian, S., Rasheed, K., Taha, T.R., and Arabnia, H.R. (2020). Automatic generation of descriptive titles for video clips using deep learning. In: *Springer Nature - Research Book Series: Transactions on Computational Science & Computational Intelligence*, Hamid R. Arabnia, Springer. 17–28.

34 Gao, L., Guo, Z., Zhang, H. et al. (2017). Video captioning with attention-based LSTM and semantic consistency. *IEEE Transactions on Multimedia* 19 (9): 2045–2055.

35 Yang, Y., Zhou, J., Ai, J. et al. (2018). Video captioning by adversarial LSTM. *IEEE Transactions on Image Processing* 27 (11): 5600–5611.

36 Singh, A., Natarajan, V., Shah, M. et al. (2019). Towards VQA models that can read. *Proceedings of the IEEE Conference on Computer Vision and Pattern Recognition*, pp. 8317–8326.

37 Jayaraman, D. and Grauman, K. (2017). Learning image representations tied to egomotion from unlabeled video. *International Journal of Computer Vision* 125 (1–3): 136–161.

38 Jayaraman, D., Gao, R., and Grauman, K. (2018). Shapecodes: self-supervised feature learning by lifting views to viewgrids. *Proceedings of the European Conference on Computer Vision (ECCV)*, pp. 120–136.

39 Gao, R., Feris, R., and Grauman, K. (2018). Learning to separate object sounds by watching unlabeled video. *Proceedings of the European Conference on Computer Vision (ECCV)*, pp. 35–53.

40 Parekh, S., Essid, S., Ozerov, A. et al. (2017). Guiding audio source separation by video object information. *2017 IEEE Workshop on Applications of Signal Processing to Audio and Acoustics (WASPAA)*, IEEE, pp. 61–65.

41 Pu, J., Panagakis, Y., Petridis, S., and Pantic, M. (2017). Audio-visual object localization and separation using low-rank and sparsity. *2017 IEEE International Conference on Acoustics, Speech and Signal Processing (ICASSP)*, IEEE, pp. 2901–2905.

42 Parekh, S., Essid, S., Ozerov, A. et al. (2017). Motion informed audio source separation. *2017 IEEE International Conference on Acoustics, Speech and Signal Processing (ICASSP)*, IEEE, pp. 6–10.

43 Asali, E., Shenavarmasouleh, F., Mohammadi, F. et al. (2020). DeepMSRF: A novel deep multimodal speaker recognition framework with feature selection. *ArXiv*, abs/2007.06809.

44 Aloimonos, J., Weiss, I., and Bandyopadhyay, A. (1988). Active vision. *International Journal of Computer Vision* 1 (4): 333–356.

45 Ballard, D.H. (1991). Animate vision. *Artificial Intelligence* 48 (1): 57–86.

46 Ballard, D.H. and Brown, C.M. (1992). Principles of animate vision. *CVGIP: Image Understanding* 56 (1): 3–21.

47 Bajcsy, R. (1988). Active perception. *Proceedings of the IEEE* 76 (8): 966–1005.

48 Roy, S.D., Chaudhury, S., and Banerjee, S. (2004). Active recognition through next view planning: a survey. *Pattern Recognition* 37 (3): 429–446.

49 Tung, H.-Y.F., Cheng, R., and Fragkiadaki, K. (2019). Learning spatial common sense with geometry-aware recurrent networks. *Proceedings of the IEEE Conference on Computer Vision and Pattern Recognition*, pp. 2595–2603.

50 Jayaraman, D. and Grauman, K. (2018). End-to-end policy learning for active visual categorization. *IEEE Transactions on Pattern Analysis and Machine Intelligence* 41 (7): 1601–1614.

51 Yang, J., Ren, Z., Xu, M. et al. (2019). Embodied visual recognition.

52 Das, A., Datta, S., Gkioxari, G. et al. (2018). Embodied question answering. *Proceedings of the IEEE Conference on Computer Vision and Pattern Recognition Workshops*, pp. 2054–2063.

53 Wijmans, E., Datta, S., Maksymets, O. et al. (2019). Embodied question answering in photorealistic environments with point cloud perception. *Proceedings of the IEEE Conference on Computer Vision and Pattern Recognition*, pp. 6659–6668.

54 Das, A., Gkioxari, G., Lee, S. et al. (2018). Neural modular control for embodied question answering. *arXiv preprint arXiv:1810.11181*.

55 Gordon, D., Kembhavi, A., Rastegari, M. et al. (2018). IQA: Visual question answering in interactive environments. *Proceedings of the IEEE Conference on Computer Vision and Pattern Recognition*, pp. 4089–4098.

56 Hinchey, M.G., Sterritt, R., and Rouff, C. (2007). Swarms and swarm intelligence. *Computer* 40 (4): 111–113.

57 Wang, J., Feng, Z., Chen, Z. et al. (2018). Bandwidth-efficient live video analytics for drones via edge computing. *2018 IEEE/ACM Symposium on Edge Computing (SEC)*, IEEE, pp. 159–173.

58 Camazine, S., Visscher, P.K., Finley, J., and Vetter, R.S. (1999). House-hunting by honey bee swarms: collective decisions and individual behaviors. *Insectes Sociaux* 46 (4): 348–360.

59 Langton, C.G. (1995). *Artificial Life: An Overview*. Cambridge, MA: MIT.

60 Hara, F. and Pfeifer, R. (2003). *Morpho-Functional Machines: The New Species: Designing Embodied Intelligence*. Springer Science & Business Media.

61 Murata, S., Kamimura, A., Kurokawa, H. et al. (2004). Self-reconfigurable robots: platforms for emerging functionality. In: *Embodied Artificial Intelligence*, (Fumiya Iida, Rolf Pfeifer, Luc Steels et al.), 312–330. Springer.

62 Steels, L. (2001). Language games for autonomous robots. *IEEE Intelligent systems* 16 (5): 16–22.

63 Steels, L. (2003). Evolving grounded communication for robots. *Trends in Cognitive Sciences* 7 (7): 308–312.

64 Durrant-Whyte, H. and Bailey, T. (2006). Simultaneous localization and mapping: Part I. *IEEE Robotics and Automation Magazine* 13 (2): 99–110.

65 Gupta, S., Davidson, J., Levine, S. et al. (2017). Cognitive mapping and planning for visual navigation. *Proceedings of the IEEE Conference on Computer Vision and Pattern Recognition*, pp. 2616–2625.

66 Zhu, Y., Mottaghi, R., Kolve, E. et al. (2017). Target-driven visual navigation in indoor scenes using deep reinforcement learning. *2017 IEEE International Conference on Robotics and Automation (ICRA)*, IEEE, pp. 3357–3364.

67 Pomerleau, D.A. (1989). Alvinn: An autonomous land vehicle in a neural network. In: *Advances in Neural Information Processing Systems*, 305–313. https://apps.dtic.mil/sti/pdfs/ADA218975.pdf.

68 Sadeghi, F. and Levine, S. (2016). CAD2RL: Real single-image flight without a single real image. *arXiv preprint arXiv:1611.04201.*

69 Wu, Y., Wu, Y., Gkioxari, G., and Tian, Y. (2018). Building generalizable agents with a realistic and rich 3D environment. *arXiv preprint arXiv:1801.02209.*

70 Kolve, E., Mottaghi, R., Han, W. et al. (2017). AI2-THOR: An interactive 3D environment for visual AI. *arXiv preprint arXiv:1712.05474.*

71 Xia, F., Zamir, A.R., He, Z. et al. (2018). Gibson Env: real-world perception for embodied agents. *Proceedings of the IEEE Conference on Computer Vision and Pattern Recognition*, pp. 9068–9079.

72 Yan, C., Misra, D., Bennnett, A. et al. (2018). CHALET: Cornell house agent learning environment. *arXiv preprint arXiv:1801.07357.*

73 Savva, M., Chang, A.X., Dosovitskiy, A. et al. (2017). MINOS: Multimodal indoor simulator for navigation in complex environments. *arXiv preprint arXiv:1712.03931.*

74 Savva, M., Kadian, A., Maksymets, O. et al. (2019). Habitat: A platform for embodied AI research. *Proceedings of the IEEE International Conference on Computer Vision*, pp. 9339–9347.

75 Datta, S., Maksymets, O., Hoffman, J. et al. (2020). Integrating egocentric localization for more realistic point-goal navigation agents. *arXiv preprint arXiv:2009.03231.*

76 Song, S., Yu, F., Zeng, A. et al. (2017). Semantic scene completion from a single depth image. *Proceedings of the IEEE Conference on Computer Vision and Pattern Recognition*, pp. 1746–1754.

77 Chang, A., Dai, A., Funkhouser, T. et al. (2017). Matterport3D: learning from RGB-D data in indoor environments. *arXiv preprint arXiv:1709.06158.*

78 Jaderberg, M., Mnih, V., Czarnecki, W.M. et al. (2016). Reinforcement learning with unsupervised auxiliary tasks. *arXiv preprint arXiv:1611.05397.*

79 Dosovitskiy, A. and Koltun, V. (2016). Learning to act by predicting the future. *arXiv preprint arXiv:1611.01779.*

80 Singh, S.P. and Sutton, R.S. (1996). Reinforcement learning with replacing eligibility traces. *Machine Learning* 22 (1–3): 123–158.

81 Schulman, J., Wolski, F., Dhariwal, P. et al. (2017). Proximal policy optimization algorithms. *arXiv preprint arXiv:1707.06347.*

82 Mur-Artal, R. and Tardos, J.D. (2017). ORB-SLAM2: An open-source SLAM system for monocular, stereo, and RGB-D cameras. *IEEE Transactions on Robotics* 33 (5): 1255–1262.

83 Mishkin, D., Dosovitskiy, A., and Koltun, V. (2019). Benchmarking classic and learned navigation in complex 3D environments. *arXiv preprint arXiv:1901.10915.*

84 Anderson, P., Chang, A., Chaplot, D.S. et al. (2018). On evaluation of embodied navigation agents. *arXiv preprint arXiv:1807.06757.*

85 Jackson, F. (1982). Epiphenomenal qualia. *The Philosophical Quarterly (1950-)* 32 (127): 127–136.

86 Tye, M. (1997). Qualia. https://seop.illc.uva.nl/entries/qualia/.

87 Hosoda, K. (2004). Robot finger design for developmental tactile interaction. In: *Embodied Artificial Intelligence*, (Fumiya Iida, Rolf Pfeifer, Luc Steels et al.), 219–230. Springer.

88 Floreano, D., Mondada, F., Perez-Uribe, A., and Roggen, D. (2004). Evolution of embodied intelligence. In: *Embodied Artificial Intelligence*, (Fumiya Iida, Rolf Pfeifer, Luc Steels et al.), 293–311. Springer.

89 Floreano, D., Husbands, P., and Nolfi, S. (2008). *Evolutionary Robotics. Technical report*. Springer-Verlag. https://infoscience.epfl.ch/record/111527.

90 Pfeifer, R. and Iida, F. (2004). Embodied artificial intelligence: trends and challenges. In: *Embodied Artificial Intelligence*, (Fumiya Iida, Rolf Pfeifer, Luc Steels et al.), 1–26. Springer.

4

Analysis of Different Regression Techniques for Battery Capacity Prediction

Param Khakhar[1] *and Rahul Kumar Dubey*[2]

[1] *Indian Institute of Technology Delhi, Department of Computer Science and Engineering, Delhi, India*
[2] *Robert Bosch Engineering and Business Solutions Ltd., Bengaluru, India*

4.1 Introduction

Lithium-ion batteries are used increasingly nowadays in a wide variety of applications. This can be attributed to the fact that lithium is the lightest of all the metals, has great electrochemical potential, and therefore provides the largest energy density for weight, which is twice that of the traditional nickel–cadmium battery. The maintenance for the Li-ion battery is low as compared with its other counterparts. However, Li-ion battery has disadvantages as well. They are structurally more fragile than their other counterparts and are prone to aging. Consequently, the batter would have reduced battery capacity. After a certain number of charge and discharge cycles, the battery would have to be replaced. Battery capacity is a measure of the amount of energy stored in the battery, and it can have the units such as ampere-hour, milliampere-hour, etc. It depends on several factors such as cycling, ambient temperature, battery chemistry, application, maintenance, etc. This chapter used machine learning algorithms to model the relation of battery capacity with various charge and discharge cycles. The widespread use of Li-ion batteries over many different scales makes the end-of-life prediction of the batteries very important as failure to replace the batteries in time may lead to adverse situations. The real-time prediction of battery capacity for Li-ion batteries would not only prevent failures but also enables to utilize the battery completely.

The advances in the computational abilities of various systems have made the use of machine learning algorithms for various prediction tasks very much feasible. This chapter analyzes the performance of several different machine learning algorithms for predicting the battery capacity of the Li-ion batteries. Several researchers have carried out experiments based on different theoretical

Cyberphysical Smart Cities Infrastructures: Optimal Operation and Intelligent Decision Making, First Edition. Edited by M. Hadi Amini and Miadreza Shafie-khah.

and analytical models. A model based on multiple concurrent degradation mechanisms was proposed by Downey et al. [1]. An approach based on normally distributed data and support vector machine (SVM) regression was studied by Susilo et al. [2]. Another experiment for the estimation of SoH and RUL of the battery system by the help of Rao-Blackwellized particle filter was carried out by Mejdoubi et al. [3]. An approach based on artificial neural network and Kalman filter for improving the health management system of the battery was proposed by Bai et al. [4]. Other filtering-based approaches used for the health management are particle filtering [5], unscented particle filtering [6], and Gauss–Hermite particle filter (GHPF) were proposed by Li et al. [7]. Brownian motion technique was used by Wang and Tsui for modeling the battery prognostics based on the drift coefficient [8].

This chapter presents an analysis of different learning algorithms used for modeling the battery capacity with respect to different charge and discharge cycles. The validation is carried out in two different ways to improve the robustness of the results obtained. This chapter is constructed with the following sections: Section 4.2 discusses the data preparation part for the experiments, Section 4.3 provides concise information about the learning models used and the experiment that is carried out, Section 4.4 discusses the results and inferences, and Section 4.5 concludes the analysis and investigates possible future continuation.

4.2 Data Preparation

4.2.1 Dataset

The dataset used for the experiments is obtained from the National Aeronautics and Space Administration (NASA) Ames Prognostic Center of Excellence (PCoE) database [9]. This dataset contains the test results of lithium-ion 1850-sized rechargeable batteries, and the experiments were performed in the NASA prognostic testbeds [10] under controlled conditions. Collected data comprises three different modes of operation, namely, charge, discharge, and impedance measurement. These modes are carried out one after another in cycles. The charging was carried out at a constant current of 1.5 A until the voltage across the battery reached 4.2 V, and then it continued charging at a constant voltage until the charge current dropped to 20 μA. The discharging process was carried out at a constant current of 2 A until the voltage dropped to 2.7, 2.5, and 2.2 V for the batteries numbered 5, 6, 7, and 18, respectively. The impedance testing was performed using electrochemical impedance spectroscopy (EIS) with frequency adjustments between 0.1 and 5 kHz. The charge and discharge cycles were performed repeatedly until the battery capacity reduced to 30% of the rated capacity.

4.2.2 Feature Extraction

For extracting features, correlation analysis was used along with data visualization. The features extracted are as follows:

(1) Average voltage across the battery during a discharging cycle (AVD).
(2) Average current through the battery during a discharging cycle (ACD).
(3) Average power delivered by the battery during a discharging cycle (APD).
(4) The time required by the battery to reach the threshold voltage for each discharging cycle (TD).

The following are the visualizations of the above features for the battery B0006. Figures 4.1–4.4 depict the average voltage, current, and power by the battery during a discharging cycle. Figure 4.5 depicts time required by the battery B0006 to reach the threshold voltage for each discharging cycle. Table 4.1 represents the correlation between the features and the battery capacity measured for each discharging cycle for the battery B0006. After extracting the above features for each of the batteries, the training data was formed. The number of instances in the training data was 168 for batteries B0005, B0006, B0007, and 132 for battery B0018. For generalizing the results, the data were scaled. The scaling was performed as follows:

- AVD was scaled by dividing the values by 4.2 V, which is the initial voltage at which discharging starts.

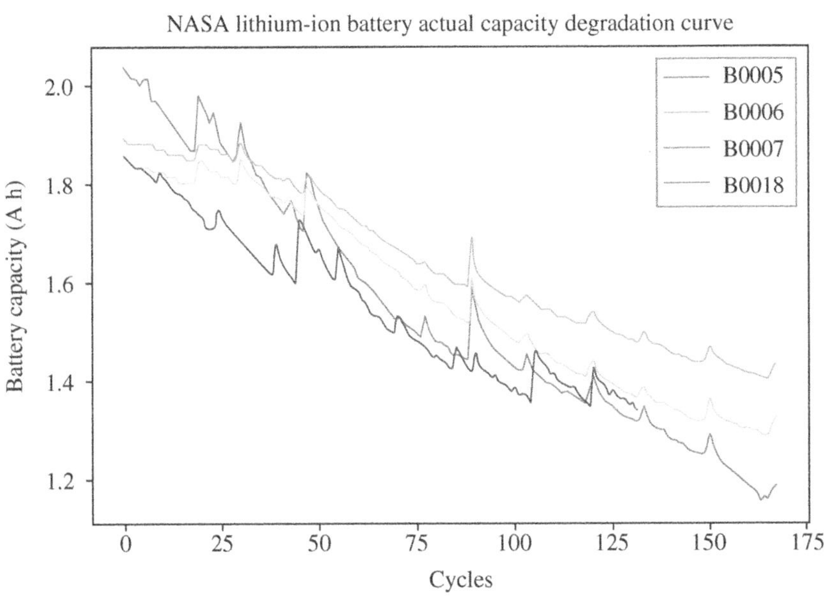

Figure 4.1 Battery capacity for different batteries with the usage of the battery.

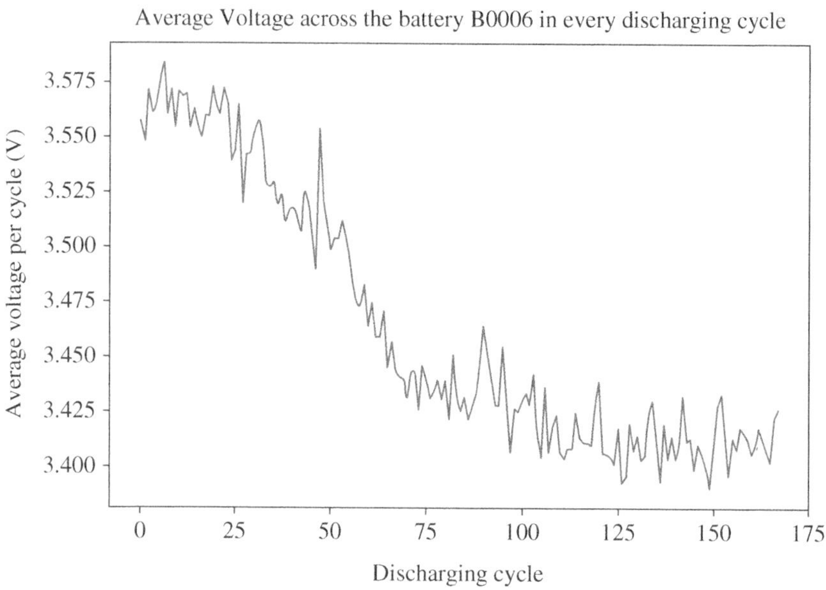

Figure 4.2 Average voltage for each discharging cycle of battery B0006 with use.

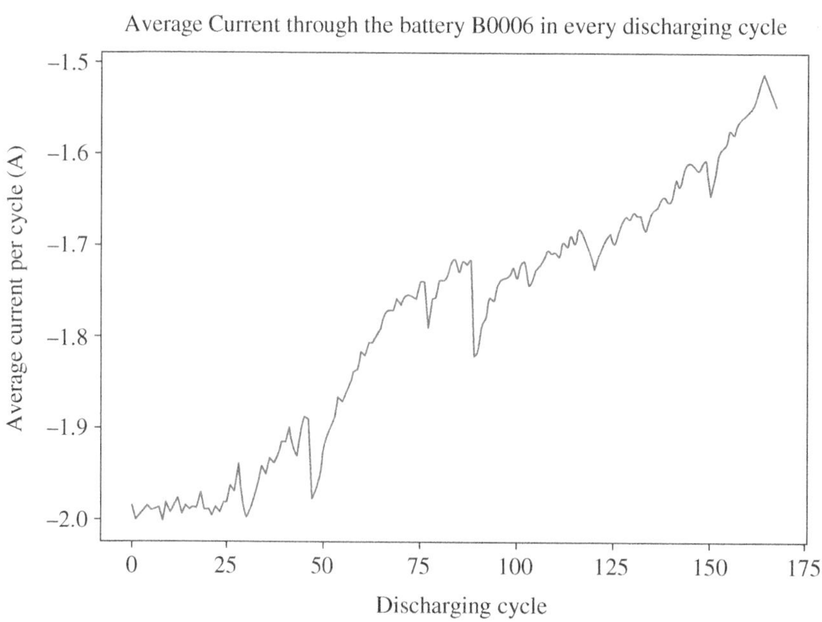

Figure 4.3 Average current for each discharging cycle of battery B0006 with use.

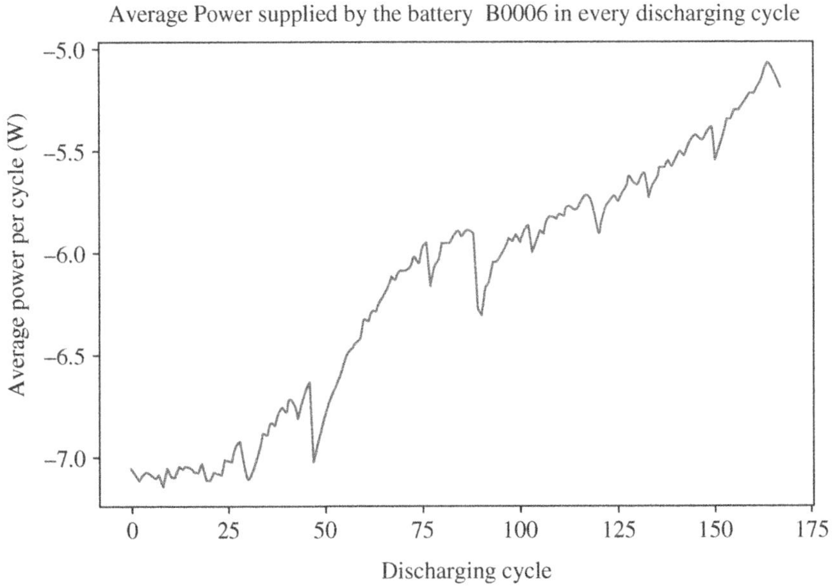

Figure 4.4 Average power for each discharging cycle of battery B0006 with use.

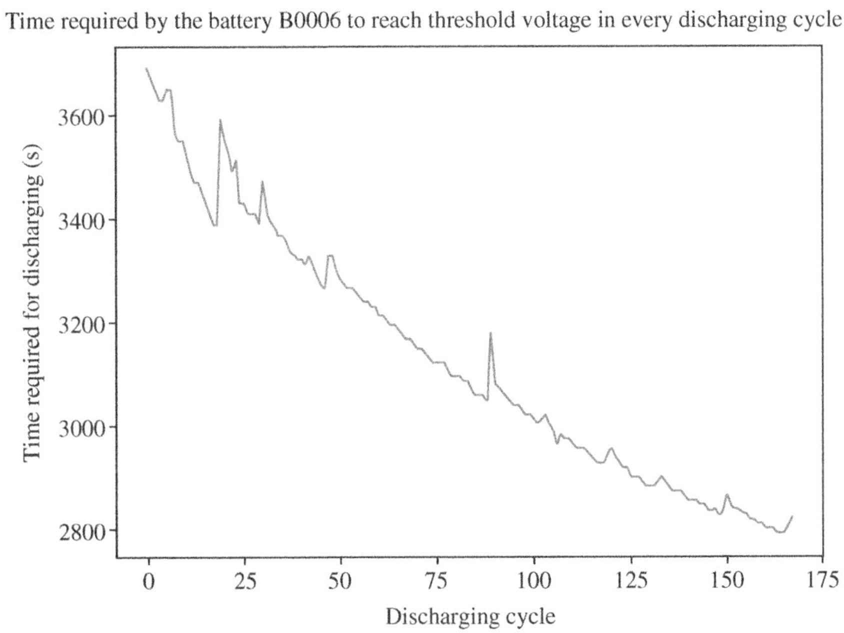

Figure 4.5 Time required by the battery B0006 to reach the threshold voltage for each discharging cycle (TD).

Table 4.1 Correlation coefficients for battery B0006.

Feature	Correlation coefficient with battery capacity
AVD	0.960
ACD	−0.989
APD	−0.992
TD	0.993

- ACD was scaled by 2 A, which is the constant current drawn from the battery.
- APD was scaled by 8.4 W, the product of constants 4.2, and 2.
- TD was scaled by the time required for the first discharging cycle for each battery.
- The battery capacity is divided by the original value 2 A h[1].

The above scaling procedure was carried out for all the batteries. Another benefit of generalization would be to use the models built in the experiment for real-time predictions of other batteries.

4.2.3 Noise Addition

The measurements of the above features could be made by sensor-based instruments. Since these instruments are prone to noisy measurements, the results calculated by intentionally adding noise to the data would provide more robust results. Gaussian noise was added to each value of the raw signal. The signal-to-noise ratio used was 30 dB. Further calculations lead to the conclusion that the noise amplitude was 0.0316 times the value of the current/voltage signal value. The noise value was then sampled from a normal distribution with 0 mean and standard deviation equal to the noise amplitude.

4.3 Experiment Design and Machine Learning Algorithms

For modeling purposes, the following machine learning algorithms were used:

(a) **Linear regression**: this models the relationship between the dependent and the independent variables. The objective is to minimize the cumulative squared error over all the samples. The weights $w = (w_1, \ldots, w_p)$ for p different

1 Ah refers to Ampere Hour.

features are changed over the training period. The loss function for the linear regression is

$$\min_{w} ||Xw - y||_2^2$$

where X is the matrix of independent values, y is the target vector, and w is the weight matrix.

(b) **Decision tree regressor:** It involves making splits in the feature space based on the Gini index for classification trees, whereas regression trees make splits based on the squared errors. In the end, a piecewise constant function is obtained, and therefore predictions are made according to that.

(c) **Random forest regressor:** this is an ensemble model of the above-discussed decision tree regressor. Different trees are trained on the bootstrapped versions of the data. While making a split, a subset of all the features is used for making a split. This ensures uniformity and prevents overfitting.

(d) **Bayesian ridge regression:** this estimates a probabilistic model of the regression problem as described above. The prior of the coefficient w is given by a spherical Gaussian:

$$p(w \mid \lambda) = \mathcal{N}(w \mid 0, \lambda^{-1}\mathbf{I}_p).$$

The priors over the regularization parameters α and γ are chosen to be gamma distributions, the conjugate prior for the precision of the Gaussian. They get estimated by maximizing the *log marginal likelihood*.

(e) **Ensemble:** in this approach, an ensemble model composing of linear regression, random forest regressor, and Bayesian ridge regression was used. The predictions from each of the algorithms were averaged to form the final prediction.

4.4 Result and Analysis

For a more comprehensive and robust analysis, there are two different validation strategies employed:

(i) The first scheme involved training and testing the learning algorithms on the data of the same battery. In this approach, the train and test data are split from the original data of a battery way such that the test data is spread uniformly. For instance, for a 75/25 split of train/test data, every fourth instance of the original dataset is included in the test. In this manner, 80/20 and 90/10 splits are considered. Every experiment was repeated 100 times to remove the effect of randomness. This was done for all batteries B0005 (Figure 4.6, Table 4.2), B0006 (Figure 4.7, Table 4.3) B0007 (Figure 4.8, Table 4.4), and B0018 (Figure 4.9, Table 4.5).

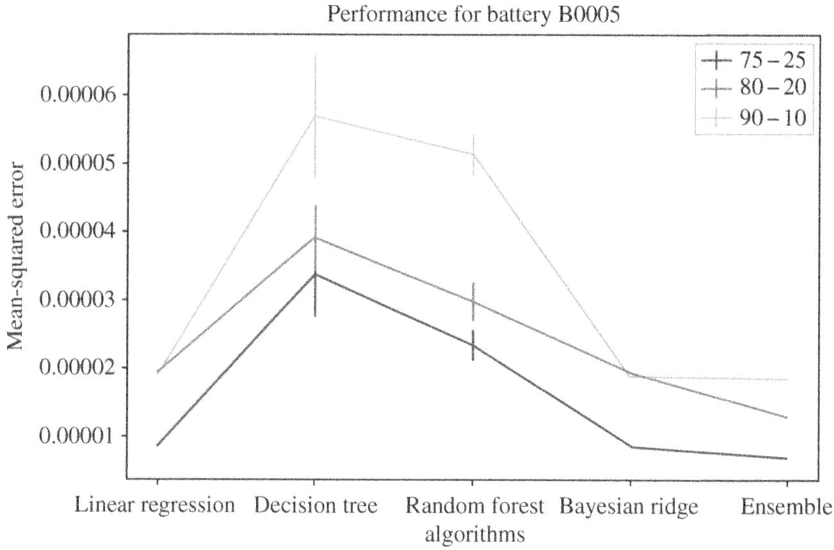

Figure 4.6 Results for battery B0005 displaying relative comparison between different learning algorithms.

Table 4.2 Results for battery B0005.

	75-25		80-20		90-10	
Learning algorithm	**MSE**	**SD**	**MSE**	**SD**	**MSE**	**SD**
Linear regression	8.64e−06	3.39e−21	1.96e−05	3.39e−21	1.90e−05	0.0
Decision tree	3.34e−05	6.35e−06	3.79e−05	5.51e−06	5.94e−05	9.4e−06
Random forest	2.39e−05	2.21e−06	3.01e−05	2.26e−06	5.09e−05	2.84e−06
Bayesian ridge	8.63e−06	1.69e−21	1.93e−05	3.39e−21	1.87e−05	0.0
Ensemble	6.95e−06	1.82e−07	1.31e−06	3.36e−07	1.87e−05	3.06e−07

(ii) The second scheme involved training the learning algorithms on the combined data of all the batteries. Data splitting is carried out as mentioned in Scheme 4.1, but the train and test data for all the batteries combined to form the final train and test sets. The train/test splits considered for the base data were again 75/25, 80/20, and 90/10. The experiments were repeated 100 times to remove the effect of randomness.

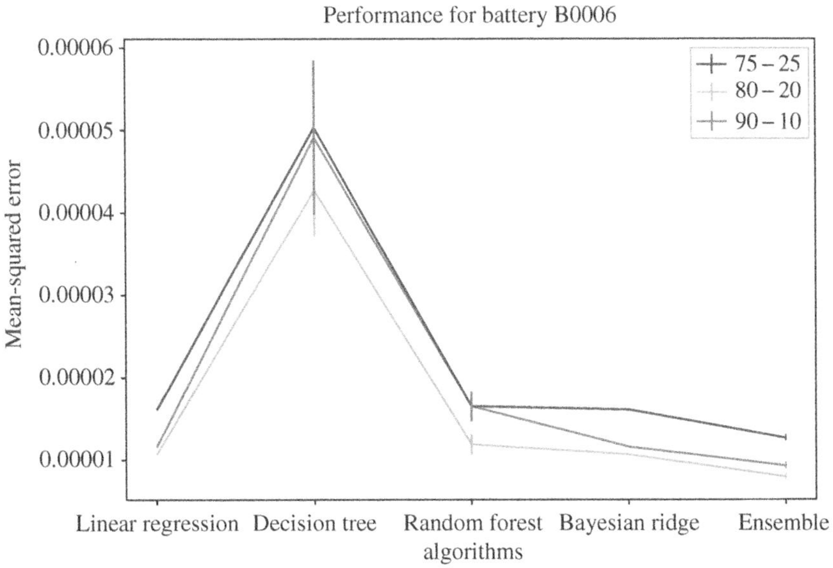

Figure 4.7 Results for battery B0006 displaying relative comparison between different learning algorithms.

Table 4.3 Results for battery B0006.

	75–25		80–20		90–10	
Learning algorithm	**MSE**	**SD**	**MSE**	**SD**	**MSE**	**SD**
Linear regression	1.61e–05	3.39e–21	1.08e–05	3.39e–21	1.15e–05	1.69e–21
Decision tree	5.07e–05	4.49e–06	4.30e–05	5.06e–06	5.02e–05	9.30e–06
Random forest	1.67e–05	1.12e–06	1.18e–05	1.06e–06	1.62e–05	1.81e–06
Bayesian ridge	1.61 e–06	0.0	1.08e–05	1.69e–21	1.15e–05	3.39e–21
Ensemble	1.27 e–06	3.04e–07	8.01e–06	2.73e–07	9.39e–06	3.79e–07

The results for the validation Scheme 4.1 for each battery are as follows:

- For the first validation scheme, the ensemble approach performs the best except for battery B0007 (Figure 4.8, Table 4.4), therefore indicating differences in the way different algorithms are modeling the data. Good results for linear regression can be attributed to the high correlation coefficient between the features. The predictions made do not fluctuate much that is evident from the low standard deviation values. The performance of Bayesian ridge regression can be attributed to the inherent structure of the data and the ability of the algorithm to model noisy measurements.

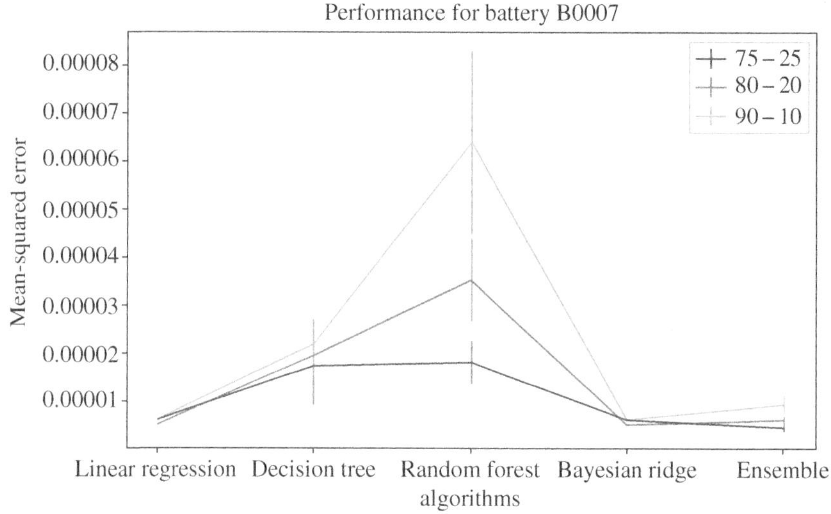

Figure 4.8 Results for battery B0007 displaying relative comparison between different learning algorithms.

Table 4.4 Results for battery B0007.

	75-25		80-20		90-10	
Learning algorithm	**MSE**	**SD**	**MSE**	**SD**	**MSE**	**SD**
Linear regression	6.02e−06	8.47e−22	5.12e−06	1.69e−21	6.05e−06	2.54e−21
Decision tree	1.71e−05	7.95e−06	1.96e−05	3.97e−06	2.17e−05	5.24e−06
Random forest	1.81e−05	4.44e−06	3.51e−05	8.54e−06	6.38e−05	1.90e−05
Bayesian ridge	6.07e−06	2.54e−21	5.03e−06	8.47e−22	5.97e−06	8.47e−22
Ensemble	4.62e−06	3.99e−07	5.87e−06	8.36e−07	9.31e−06	1.81e−06

The second validation strategy involves the training models on three out of the four datasets at a time, and the unused dataset is used as the test set. The experiment *i* performed 100 times for all the 4 datasets, and the final mean squared error is obtained after averaging the results of all the experiments. The results obtained along with the comparison between different methods are as follows:

- For the second validation scheme, the predictions of all the algorithms have higher mean squared error (Figure 4.10, Table 4.6), thus indicating the difference in the internal structures of the battery. The fluctuations are less for all the algorithms except decision tree regressor. Tree-based methods made predictions with lower mean squared error.

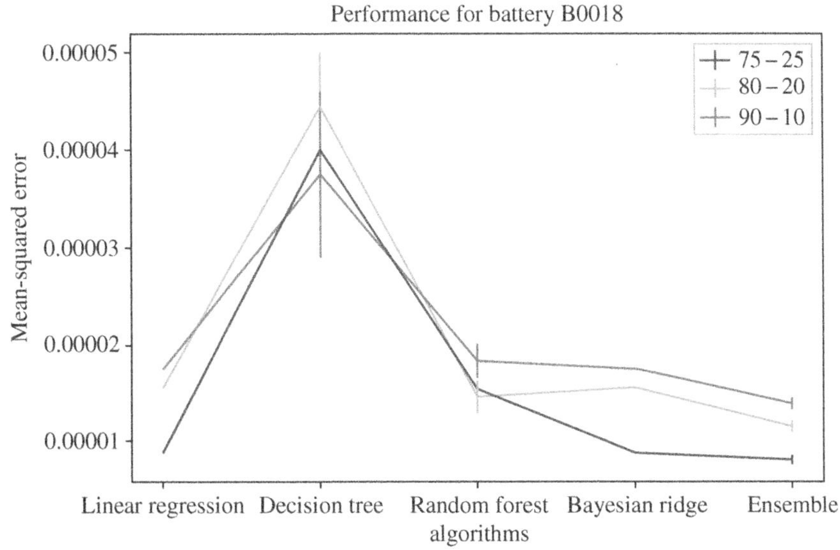

Figure 4.9 Results for battery B0018 displaying relative comparison between different learning algorithms.

Table 4.5 Results for battery B0018.

	75–25		80–20		90–10	
Learning algorithm	**MSE**	**SD**	**MSE**	**SD**	**MSE**	**SD**
Linear regression	8.75e–06	3.39e–21	1.55e–05	0.0	1.74e–05	0.0
Decision tree	3.98e–05	3.49e–06	4.40e–05	5.06e–06	3.76e–05	1.04e–05
Random forest	1.52e–05	9.80e–07	1.49e–05	1.41e–06	1.81e–05	1.63e–06
Bayesian ridge	8.64e–06	1.69e–21	1.55e–05	3.39e–21	1.74e–05	3.39e–21
Ensemble	8.07e–06	1.90e–07	1.14e–05	6.74e–07	1.38e–05	5.32e–07

4.5 Threats to Validity

The data used for the experiments was created by repeated experimentation by researchers at NASA. The real-life usage of batteries may differ significantly from the experiment carried out at NASA, and therefore, there might be changes in the internal constituents of the battery resulting in different trends. The ambient temperature was kept constant during the data collection at NASA, which would not be the case in a real-life setting. For modeling the algorithms in this

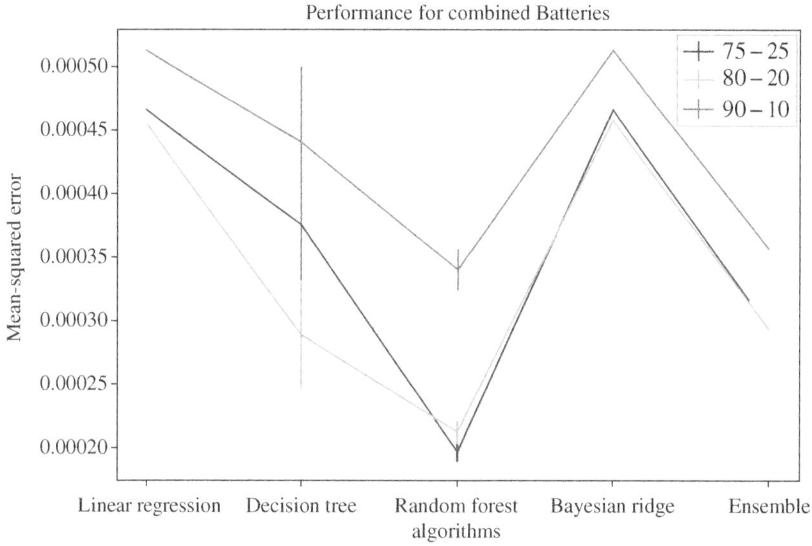

Figure 4.10 Results for combined batteries displaying relative comparison between different learning algorithms.

Table 4.6 Results for combined batteries.

	75-25		80-20		90-10	
Learning algorithm	**MSE**	**SD**	**MSE**	**SD**	**MSE**	**SD**
Linear regression	4.67 e−04	5.42e−20	4.54e−04	1.08e−19	5.13e−04	0.0
Decision tree	3.76e−04	6.19e−05	2.89e−04	4.30e−05	4.41e−04	5.91e−05
Random forest	1.98e−04	9.80e−07	2.11e−04	8.94e−06	3.40e−04	1.66e−06
Bayesian ridge	4.61e−04	1.69e−21	4.57e−04	5.42e−20	5.13e−04	0.0
Ensemble	2.94e−04	1.90e−07	2.94e−04	2.84e−06	3.57e−04	3.67e−06

chapter, the temperature was excluded as a factor influencing the battery capacity that might not be the case as well. The data preprocessing techniques employed in this chapter are used for generalizing this to all the batteries, but there can be certain other physical, structural, and ambient factors that might influence the battery capacity.

4.6 Conclusions

This chapter analyzed the performance of various regression algorithms for the task of predicting battery capacity in advance. Relevant feature extraction and feature engineering were employed, and then random noise was added to the data. Two different validation schemes were used for making the results more comprehensive, robust, and generalized. The models can be used for real-time prediction of battery capacity as the training errors are in the order of 1e−06 (for first validation scheme); therefore the predictions are correct up to three decimal digits that is very reasonable.

References

1 Downey, A., Lui, Y.H., Hu, C. et al. (2019). Physics-based prognostics of lithium-ion battery using non-linear least squares with dynamic bounds. *Reliability Engineering and System Safety* 182: 1–12.

2 Susilo, D.D., Widodo, A., Prahasto, T., and Nizam, M. (2018). State of health estimation of lithium-ion batteries based on combination of Gaussian distribution data and least squares support vector machines regression. In: *Materials Science Forum*, vol. 929, 93–102. Princeton, NJ: Trans Tech Publications.

3 El Mejdoubi, A., Chaoui, H., Gualous, H. et al. (2019). Lithium-ion batteries health prognosis considering aging conditions. *IEEE Transactions on Power Electronics* 34 (7): 6834–6844. https://doi.org/10.1109/TPEL.2018.2873247.

4 Bai, G., Wang, P., Hu, C., and Pecht, M. (2014). A generic model-free approach for lithium-ion battery health management. *Applied Energy* 135: 247–260.

5 Saha, B. and Goebel, K. (2009). Modeling Li-ion battery capacity depletion in a particle filtering framework. In: *Proceedings of the Annual Conference of the Prognostics and Health Management Society, San Diego, CA, USA*, 2909–2924.

6 Miao, Q., Xie, L., Cui, H. et al. (2013). Remaining useful life prediction of lithium-ion battery with unscented particle filter technique. *Microelectronics and Reliability* 53: 805–810.

7 Li, B., Peng, K., and Li, G. (2018). State-of-charge estimation for the lithium-ion battery using the Gauss-Hermite particle filter technique. *Journal of Renewable and Sustainable Energy* 10: 014105.

8 Wang, D. and Tsui, K.L. (2018). Brownian motion with adaptive drift for remaining useful life prediction: Revisited. *Mechanical Systems and Signal Processing* 99: 691–701.

9 Saha, B. and Goebel, K. (2007). *Battery Data Set, NASA Ames Prognostics Data Repository*. Moffett Field, CA: NASA Ames http://ti.arc.nasa.gov/project/prognostic-datarepository.

10 Saha, B. and Goebel, K. (2008). Uncertainty management for diagnostics and prognostics of batteries using Bayesian techniques. In: *Proceedings of the Aerospace Conference, Big Sky, MT, USA*, 1–8.

5

Smart Charging and Operation of Electric Fleet Vehicles in a Smart City

Milad Kazemi[1], Samuel Bailey[1], Sadegh Soudjani[1], and Vahid Vahidinasab[2]

[1] *School of Computing, Newcastle University, Newcastle upon Tyne, UK*
[2] *Department of Engineering, School of Science and Technology, Nottingham Trent University, Nottingham, UK*

5.1 Smart Charging in Transportation

In this chapter, we discuss a number of important areas relating to the electric vehicles (EVs), EV smart charging, and interaction of smart charging with the smart grid. First, we explore the technologies currently available for charging EVs and the rules and regulations in place in countries around the world. Then, we discuss the cyber-physical aspects of the smart charging system in Section 5.2. Security is essential in the modern age, with a number of attack methods and possible countermeasures discussed in Section 5.4. Finally, we study a number of proposed smart charging schemes, each with its own purpose, benefits, and drawbacks in Section 5.5, and provide a numerical example of computing a smart charging scheme using techniques from game theory.

5.1.1 Available EV Charging Technologies

5.1.1.1 Inductive Charging

Inductive charging is a technology being explored for easy, autonomous charging of EVs, and it is a technology already widely used in mobile phones and is seeing use in various countries for EV charging [1, 2]. BMW is the first automobile manufacturer offering wireless charging with their vehicles [3]. Other companies such as WiTricity, HEVO, and Plugless are also working on wireless EV charging technologies. The Society of Automotive Engineers (SAE) recently published the SAE J2954 standard [4] for EV charging and the SAE J2847/6 recommended practice [5] for the communication between EVs and charging pads.

Cyberphysical Smart Cities Infrastructures: Optimal Operation and Intelligent Decision Making, First Edition. Edited by M. Hadi Amini and Miadreza Shafie-khah.

5.1.1.2 Battery Swapping

Battery swapping or swap charging (SC) is a technology developed for easy and quick removal and replacement of a depleted onboard battery for a charged one. The most obvious benefit to SC is the quick "refueling" times. Users are able to go from a fully depleted battery to a fully charged one in a matter of a few minutes, approaching the amount of time it takes to fill a car with conventional liquid fuels.

One such example of this system is from the Chinese EV manufacturer NIO. Users access the service via a mobile app to find battery swapping stations and to reserve a slot. Information such as available batteries and remaining charge time are displayed. The whole battery swap takes approximately three minutes. Owners of new NIO vehicles are offered this service free of charge, whereas owners of used NIO vehicles can use the service by paying for the energy difference between the battery being taken out and the battery being swapped in.

5.1.1.3 Automatic Robotic Charging Connector

Automatic robotic connectors is an area of autonomous charging being explored. They work by automatically connecting a charger cable from a charging station to an EV without the intervention of a human. There are variants on this theme. One variant uses the conventional charging connector located on an EV. A robotic arm detects the position of the connector and guides the charging cable into the socket. One such implementation is Volkswagen's e-smartConnect system that uses a robotic arm to connect a conventional charging cable to an EV [6].

5.1.1.4 Automatic Ground-Based Docking Connector

Another approach is to have a charging connector located on the underside of the car. An arm built into the parking surface rises up to make contact with the connector to commence charging. This style of system allows high-power fast charging and could be retrofitted to add support for most EVs. It also works at full power without having the car and dock lined up perfectly, unlike inductive charging, as the dock can move its port to line up with the vehicles charging connector.

This system offers similar benefits as static inductive charging while being cheaper. Such a system is offered by Volterio.

5.1.2 Current Regulations on Smart Charging

With the current problem of global warming, action needs to be taken to slow down, and eventually stop, the damage to the atmosphere. This is a global concern, which will require the commitment of the world to overcome. Carbon dioxide is the worst pollutant offender [7], with transport and electricity generation producing the most CO_2 [8]. It is evident that EVs are key to a sustainable future. However, surveys have shown that potential EV customers are still concerned about factors such as cost, range anxiety, and number of charging points [9]. There are also issues with the substantial increase in the number of EVs causing instability in the current electrical grid infrastructure. Regulations imposed by councils and governments can help with this growing number of EVs by both incentivizing individuals and companies to adopt the new technology and enforcing appropriate regulations.

Net-zero carbon emission is the balancing of carbon emission and the extraction of carbon from the atmosphere. To achieve net zero, the amount of produced carbon emissions needs to equal the amount of carbon emissions absorbed from the atmosphere. Reaching net zero is critical for curbing global warming. A number of countries have passed acts that put this goal into their law. In 2019, the United Kingdom was the first of the world's major economies to put the net-zero goal into legislation, requiring the country to achieve net zero by 2050 [10]. France, Denmark, Hungary, New Zealand, and Sweden have also put net zero into law [11–15].

Tax credits and other incentives are used by countries to encourage the purchasing on new EVs by consumers. In the United States, tax credit is offered to eligible customers.[1] Out of 28 European Union member states, 27 are offering various benefits and incentives for purchasing and running EVs [19]. China is the largest market of EVs [20], as can be seen in Figure 5.1, and has extended its EV subsidy scheme from the end of 2020 to the end of 2022, although the size of the subsidy has been decreased. Figure 5.1 shows the EV stock in main markets including separate indication of plug-in hybrid electric vehicles (PHEVs) and battery electric vehicles (BEVs).

The most effective form of enforcing the move to the new technology seems to be the ban on sales of new internal combustion engine (ICE) vehicles. In 2018, the UK government announced that it would ban the sale of new ICE cars from 2040 [22]. Recently, it was confirmed that this date would be brought forward by a decade, to 2030 [23]. While pushing consumers who purchase new vehicles to switch to EV will dramatically increase the uptake of EVs, the infrastructure needs to be ready to cope with the demand. The UK government, as part of the "Automated and Electric Vehicles Act 2018" [24] (AEV), has put laws into place to ensure this goal is met.

Charger availability is a huge problem for EV owners, as the lack of readily accessible charging stations may limit how EVs can be used [25]. As part of the UK's AEV Act, the charging infrastructure must improve. This can include forcing large fuel retailers to provide charging stations. It also includes regulation on

1 In the United States, customers are eligible for up to US$ 7500 of tax credits for the first 200 000 vehicles sold by a manufacturer [16] (Tesla and General Motors have reached this cap, with their incentives now phased out [17]). There is also a 30% tax credit on residential charger and installation costs, up to US$ 1000 [18].

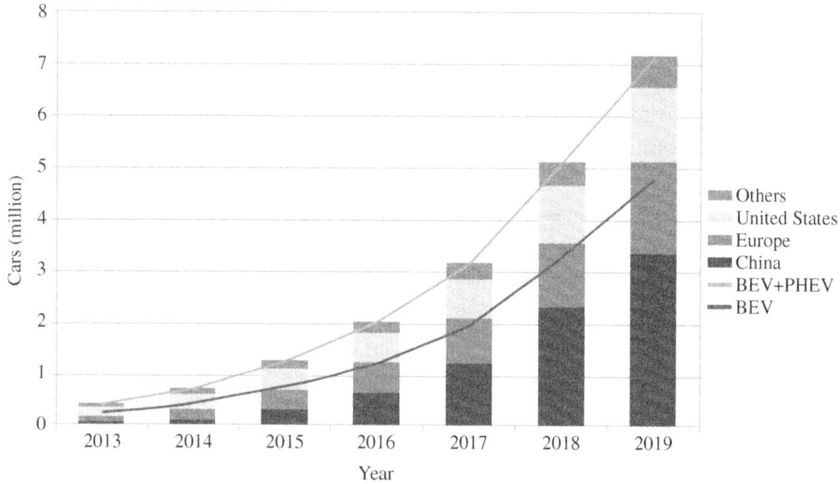

Figure 5.1 Electric car stock in main markets. Source: Data obtained from [21].

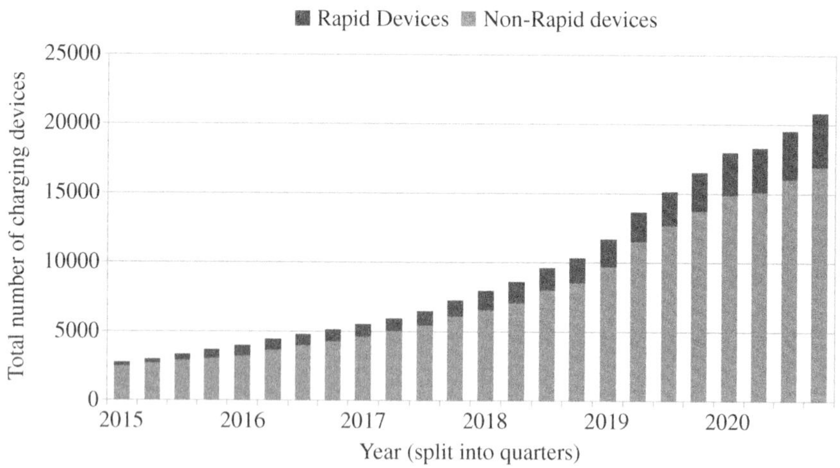

Figure 5.2 Number of charging devices in the United Kingdom. Source: Data obtained from [26].

installation of charging points for new and refurbished properties (both residential and commercial).

The number of charging points has been increasing rapidly in the United Kingdom. Figure 5.2 shows the number of public charging points by speed [27]: in 2011 being just over 1300, in 2017 just under 15 000, and in 2020 over 35 000.

Smart charging is another area covered by the UK's AEV Act. Sales of non-smart-ready chargers can be banned [24], ensuring that the chargers in use are capable of monitoring charging, changing the charging power level, and reporting this data to be used in smart charging coordination. Importantly, it also allows the enforcement of charging station operations to cooperate with each other [24], including the sharing of information, which will be vital for smart charging to work optimally.

Interoperability is an important factor in smart charging. Currently, most consumers receive a home charging station from the EV manufacturer, which can be integrated into their home's electricity distribution point to deliver low-power charging to their EV at home. If a consumer purchases a charger from their smart services provider, the charger should be interoperable, allowing the consumer to move to another provider, without losing functionality from their device. This would also exclude EV manufacturers from partnering with a smart service provider to provider locked-down chargers. Because of this, competition will remain high among providers, helping to lower costs for consumers.

5.2 Cyber-Physical Aspects of EV Networks

Smart charging can be considered as a part of a cyber-physical energy system (CPES). The CPES represent a class of widely distributed and globally interconnected energy systems that integrate *computation* processes, *communication* elements, and *control* processes. Research related to CPES include [28–32]. The future smart grids will rely on data capture and smart control of devices to deliver power, when and where it is required. The data captured from devices such as smart chargers will be used to compute optimal charging schedules for the connected EVs, which will be sent to the chargers to control the amount of power being consumed. In this section, we explain how smart charging fits into CPES.

5.2.1 Sensing and Cooperative Data Collection

Any smart system needs to be able to interact with its environment. Sensors will be vital for the safe an efficient operation of a smart grid and for the optimal charging of EVs connected to the grid. Smart chargers can be used to collect data on the state of charge (SoC) of the connected EV and control the power output to each EV in the CPES. Sensors spread around the smart grid, such as local distribution transformers, will be able to report data on the local grid load. Power generation plants will also have sensors detecting the amount of power being output and controllers that are able to regulate the production of power to meet current demand. The number of sensors in the network needs to be enough to satisfy an important feature of CPES, which is its reliability. This is critical for the CPES to be able to operate autonomously, even under device malfunction.

Vehicular ad hoc networks have recently been introduced as a technology that provides broadband services to vehicles [33]. In this network, vehicles are equipped with onboard communication devices, and wireless access points are deployed, such as roadside units (RSU) along motorways and roads. This new technology enables vehicles to communicate with each other (vehicle-to-vehicle communication) and with the RSU (vehicle-to-RSU communication).

Multiple network structures for communication can form based on this new paradigm, and each one of them has its own pros and cons. For instance, the sharing and cooperation of data collection will be important for the most efficient use of the available resources. In this case, if we use a distributed communication network and if separate charging station operators fail to share adequate data, it could lead to a situation where the grid is near full capacity due to a number of charging stations being near full capacity, but they are not able to accept any future EVs as there is no way for them to tell if there will be spare capacity later on.

The type of data required depends on optimization methods being used. Most optimization methods require the user to input their required SoC and their departure time, usually into a mobile app, which allows the optimization algorithm to shuffle the charging schedules of different users and allocate them a charging time that fits their needs but also allows other EVs to achieve their required charging state too. Optimizing ahead of time (e.g. day ahead) requires more details about when and where each EV is going to be available for charging. For V2G activity, exact SoC levels for each EV needs to be known for the coordinator to assess how much extra power is available to them at any given moment. However, this approach is not completely realistic: most of the time, the charging station does not have access to the distribution of SoC prior to EVs sharing their data with the charging station. On the other hand, if we want to use real-time optimization approaches, the limited computational power restricts the amount of data used for finding the optimal charging strategy.

A CPES should also be able to take advantage of existing infrastructure. For instance, traffic flow can be used in the prediction of future EV demand that can be utilized when designing an EV charging schedule [33]. Weather also plays a factor in charging, as colder weather means more use of the heater and hotter weather means more use of air conditioning, which will increase the power consumption of an EV, therefore requiring it to be charged more frequently. Holiday periods may also have an impact on charging, as driving becomes less regular as people are not attending work or school at set times.

Many optimization techniques consider a *centralized* view of the problem. This is where a central coordinator collects all of the charging data, EV data, and grid-related data, to come up with a charging schedule for the EVs that is then sent by the central coordinator to the network. This means all data from sensors distributed in the grid need to be sent to a central location. Due to the amount of data required, this could cause bandwidth or delay issues. An alternative approach is *distributed optimization*, requiring less data to be transmitted as each EV takes more responsibility in determining the optimum time to charge. Both of these approaches are discussed in Section 5.2.2.

5.2.2 Data-Driven Control and Optimization

Data will be key for the smart control of EV charging. Without accurate data on EVs requesting to charge and on the available capacity in the grid, it is easy for the grid to become oversubscribed or for EVs to miss charging opportunities. Charging of EVs will need to be optimized according to the data available to the charging coordinator, whether the charging organization is performed centrally or in a distributed manner. On the other hand, assuming we have full access to the data of the vehicles, we also need to take into account the driver's preferences. Moreover, because of the integration of green energy resources, we need to consider various sources of uncertainty in our analysis [34]. There are a number of current issues with control and optimization in CPESs, some of which are covered in [35].

Real-time communication: It plays a key role in a smart charging CPES. Electrical grids require quick responses to environmental changes, such as sudden increases in demand, to maintain grid stability. With the dramatic changes in loads that a large population of EVs can cause, it is vital that the design of the communication layer is able to keep up with the flow of data, ensuring it is able to reach the control applications in the shortest possible time. For instance, due to the high EV mobility and the nature of short-range transmissions, this can lead to a transmission delay and result in additional distance traveled by the EV while they are waiting the decision maker's response [33].

Diverse network architecture: Smart grids cover a wide area, meaning the distance data has to travel is large. Wireless communication already exists in such technologies as Wi-Fi and cellular connections. A diverse network architecture will be required to meet the demands of smart charging: from rural communication that requires low density but long-range data transfer to cities that have a large population of already densely packed wireless communication technologies, possibly causing interference issues for critical grid data communication.

Security: In a CPES, data needs to be shared to achieve efficient grid operation. This has potential major consequences for the privacy of the users of smart charging, as discussed in Section 5.4. Current protocols are designed mainly for functionality, rather than security, which is also discussed in Section 5.4.1. In [36], the authors consider attack scenarios that can take place in a CPES.

Domain optimization: Smart charging involves different actors as introduced in Section 5.3.1. These include energy suppliers, transmission supply operators, distribution supply operators, and energy consumers (EV owners). Each of these individuals may have a different optimization goal. For instance, an energy consumer charging their EV may want to minimize the total cost of charging their vehicle, whereas a distribution supply operator may want to optimize charging to stabilize voltage on their power distribution network. A transmission supply operator may primarily be concerned about the stability of the frequency of their grid.

Physical attacks: The physical components of a CPES can come under attack. Charging stations can be tampered with, which can pose a danger to people trying to operate the equipment, the EV being charged, and the grid itself. In [37], a scalable, tamper-resistant mutual authentication framework is described, providing authentication between the EV and charging station. Payment security is also of concern, as the charging stations may offer payment services to potential customers, which could be attacked. A possible solution would be to enforce the use of mobile app payments; however not all customers may have a smartphone capable of making such transactions.

5.3 Charging of Electric Fleet Vehicles in Smart Cities

Charging a large number of EVs poses a number of problems, as it puts a large strain on the whole electrical grid. However, a large number of EVs presents an opportunity to the grid, as a large distributed network of high-power batteries would be able to offer a number of services to the grid. In this section, we discuss intelligent management of fleets of electric vehicles and the support services they can provide.

5.3.1 Intelligent Management of Fleets of Electric Vehicles

5.3.1.1 Charging of EV Fleets

Fleet operators (FOs) will be the ones orchestrating the charging of EVs and will have a crucial role in supporting the grid operator for running the grid safely while serving the needs of the EV owners (i.e. the FO customers) and also providing services back to the different levels of the grid. This could include energy storage for renewable energy providers, frequency regulation for transmission service operators, and local flexibility and voltage regulation services for the distribution system operators (DSOs).

The electricity system operators (ESOs) are required to keep the grid frequency in a specified range at all the time. During the normal situation, they use turbine governors to ramp up and down the available power to match current demand. But in time of contingencies, where there is a strong need to fast response, EVs are a very suitable solution for frequency regulation [38, 39].

DSOs are also required to keep local power distribution networks (e.g. residential and commercial sectors of a city) working optimally [40]. EVs can be used to match demand, keeping power usage within manageable limits, dropping peak loads, and creating a steady draw from the grid [41].

Renewable energy suppliers (RESs) provide a clean but intermittent energy source that calls for more flexible resources in the grid operation. There can be times of great demand and little power from renewable sources, meaning more conventional energy, such as fossil fuel, needs to be used to meet demands. Similarly, there can be times of little demand but an abundance of renewable energy that ends up going to waste. EVs can be used as a large energy store. In times of little demand and great supply, energy can be stored in EV batteries. Then when demand is high but supply is low, EVs can supply energy back to the grid to meet the demand without resorting to more conventional means of energy generation [42, 43].

Fleet charging will consist of balancing the needs of all parties FOs will have to deliver charging when EV owners require it while also offering incentives to the owners to leverage their EVs for facilitating the requirements of ESOs, DSOs, and RESs. As discussed in [44], this has to be carefully coordinated however. The needs of one operator may conflict with the needs of another, requiring careful balance between all parties.

5.3.1.2 Route Mapping with Charging

EVs do not yet have the range of ICE vehicles, with the average range of EVs being 181 miles [45] and 412 miles for ICE vehicles [46]. In cold weather, it has been found that EV range can drop significantly, due to the cooler temperature affecting the battery and also the extra load of the heater (with ICE vehicles, heat is a by-product of combustion and does not increase load on the engine). In addition to this, the current fast charge technology still does not provide as quick energy replenishment as filling a tank with liquid fuel. For journeys that approach the maximum range of an EV, en route charging needs to be taken into account. This is especially the case in places where charging stations are sparse, as leaving charging until later on may leave the EV stranded, in which, unlike with an ICE vehicle for which it is easy to transport the energy source, there is currently little option other than the vehicle being transported to a charging station.

In [47], the authors propose a solution where routes are planned according to the remaining EV charge and the available charging stations along the possible routes. Two main objectives are considered: total journey time and money savings. Solutions are computed, and the user is shown two of the best results. The user selects their desired route, and reservations are made at the charging stations along that route. In future, it would be interesting to consider the changes of charge state on the fly, adjusting the route if required to optimize the users chosen objective, as factors such as weather and traffic can change driving conditions and energy consumption.

As the range of EVs increases, the requirement for en route charging will decrease. However, for always on the go vehicles such as fully autonomous taxis, being able to charge fully overnight may not be an option, meaning route optimization could remain important. In addition, fully autonomous methods of initiating charging will need to be devised to allow autonomous vehicles to continue operating without a human operator, even when charging en route, with automatic charging methods possibly being a consideration in future route mapping solutions.

5.3.2 Electricity Grid Support Services

While EVs will add a large load to the electrical grids around the world, they can also offer new opportunities and services to the grids. With large onboard batteries, EVs can form a large distributed energy storage network, capable of offering a number of services to the grid that will help realize the carbon-neutral goal, forming part of the infrastructure itself. These services range from demand response to shave peaks in demand to Vehicle to Grid or Vehicle to Building (V2B) for valley filling with renewable energy sources. Such services can help lower the costs for the infrastructure required for smart grids that will be powering the smart cities of the future.

5.3.2.1 Demand Response

Electricity demand varies depending on the time of day, day of the week, and seasonal. Energy demand often peaks in the morning, as people are getting up and going to work, and in the evening, when people arrive back from work. Demand response programs are to encourage the charging of EVs to react and adapt to the load on the grid. As demand from non-EV appliances increases, EV charging can be slowed down, or even stopped, to keep the grid operating safely and optimally. At times when demand is low, such as certain periods at night, EV charging can ramp up, taking advantage of the extra power available. This keeps the peak demands on the grid lower, lowering the cost of the infrastructure while still fully utilizing the grid's available capacity.

5.3.2.2 Frequency Response

Using Vehicle-to-Grid technology, EVs can supply energy from their batteries back to the grid. This would be useful in a number of scenarios, one being frequency regulation. As the load on the grid increases, the generators slow down as they come under greater load, which lowers the frequency of the grid. This is particularly apparent when there is a loss in supply. Energy suppliers have reserves in place to help deal with these situations. EVs have the ability to fill in the gap, supplying instant and large amounts of power back to the grid, minimizing impact, and lowering operator costs [48]. The authors of [39] propose a control approach using EVs to maintain frequency response under a large power outage.

5.3.2.3 Emergency Power

If total loss of power occurs, the energy stored in EV batteries can be fed back to the grid with V2G, keeping critical systems fed with power until the grid comes back online. Street lighting, hospitals, and other high-priority services can be kept running with minimal interruption. Individual EVs could also supply power to homes and buildings under blackout (Vehicle to House and V2B) or supplement personal renewable energy (e.g. rooftop solar photovoltaics [PVs]).

5.3.2.4 Emergency Response

Another opportunity for EVs to aid in an emergency could be emergency response. If some sort of local disaster occurs, EVs can be driven to the location and used to supply power to emergency facilities, such as field hospitals, until a more permanent source of energy can be set up.

5.4 Data and Cyber Security of EV Networks

Smart charging is a powerful tool and will be fundamental for supporting the energy systems inside the smart cities. However, connecting many devices together to coordinate and control EV charging brings privacy concerns, as potentially sensitive information will have to be transmitted and received [49]. This data might include a person's schedule (not directly but it can be extracted) that could be used to determine the optimum times for their EV to charge. With many people's main mode of transport being a car, bringing down the infrastructure that allows the EVs to operate would cause widespread disruption. For smart grids in general, it is possible to destroy equipment using cyber-attacks [50, 51]. Smart devices, such as smart chargers, are required to alter charging schedules of EVs and power available on the network. The increase in reliance of connected smart devices opens the attack surface for the system, opening up new methods of attack.

5.4.1 Attack Schemes

Smart devices, like any piece of technology, are susceptible to attack. Smart grids and optimized EV charging depend on the use of smart devices, such as voltage or frequency sensors, to detect, collect, and report data to charging optimization controllers.

Current industry protocols for cyber-physical systems (CPSs) are not secure, as they are largely outdated and were written with the design goal of usability and speed, rather than security [52]. Security protocols can be hard to prove and verify [53]. They are also expensive in terms of compute requirements, with very low-power microcontrollers, which might be found in chargers and sensors throughout the network, not having the resources to perform modern cryptography algorithms. This weakens the security of such systems, leaving them open to a number of attack schemes.

5.4.1.1 Data Injection

One way attackers can disrupt the system is by injecting false data in the network (false data injection [FDI] attack). This can cause false readings to be taken, causing control decisions to be made that may be detrimental to the grid. Sensors throughout the grid stream data about current demand, frequency, voltage, etc. EV charging stations transmit charging requests and also stream charging data for EVs currently being charged. Accurate data is critical to the functioning of the smart charging system.

Data injection attacks aim to affect the smart charging management system by injecting false data into the grid. For instance, if attackers inject charging readings that state there are many EVs charging at full power, the control schemes may decide to lower charging of other EVs on the network, causing major disruption. It may also be possible to target individuals using this method of attack, causing their charger unit to operate in a degraded state. A list of suggested data types can be found in [54].

5.4.1.2 Distributed Denial of Service

Attacks can be launched that aim to degrade services, disrupting communication and transmission of sensing data. Such attacks are already widely known today, with Amazon Web Services receiving a distributed denial-of-service (DDoS) attack in February 2020 [55] and GitHub in February 2018 [56]. These attacks shut down services, stopping genuine users from accessing them. As the attacks are distributed, it can be difficult to track and trace the sources to stop and prevent further attacks. Some smart charging strategies use a centralized approach with computation being performed in one place, as discussed in Section 5.3, which could be vulnerable to the same attacks.

5.4.1.3 Data and Identity Theft

Online accounts could be used to schedule and track charging and for billing purposes. This means data about energy consumption, SoC, location, planned schedules, etc. would have to be transmitted and stored. Should an attacker gain access to this information, they may be able to use it against the individuals. For instance, should an attacker learn of the location, SoC, and schedule, they may be able to determine whether or not someone is home, and for how long. Identity theft is another area of concern. One possible way to prevent attacks is to require all communicating parties to identify themselves and authenticate transmissions, preventing both spoofing of data from a persons account and repudiation attacks. For example, monitoring energy usage to bill a customer will require an identity to be attached to a message. Should an attacker manage to impersonate someone else, they may be able to gain access to their online account and start consuming electricity under their identity. This is similar to how attackers use other people's identities to gain access to accounts online that they can then use for fraudulent activity such as purchasing items using the victim's bank details.

5.4.1.4 Man-in-the-Middle Attack

A man-in-the-middle (MitM) attack is when two parties are communicating to each other. However, they have been convinced that a third party is the other participant, meaning all communication goes through the third party unbeknown to the two participants, and the third party is free to listen and modify messages, as shown in Figure 5.3. Smart grids rely on the data streamed from sensors located at different parts of the network in real time. Smart charging algorithms require accurate data on current available power and current power being used. Current implementations are often legacy systems, with more focus on utility, rather than security (for instance, the authors of [52] discuss some protocols used in smart grids, such as distributed network protocol 3 (DNP3) and modbus, which

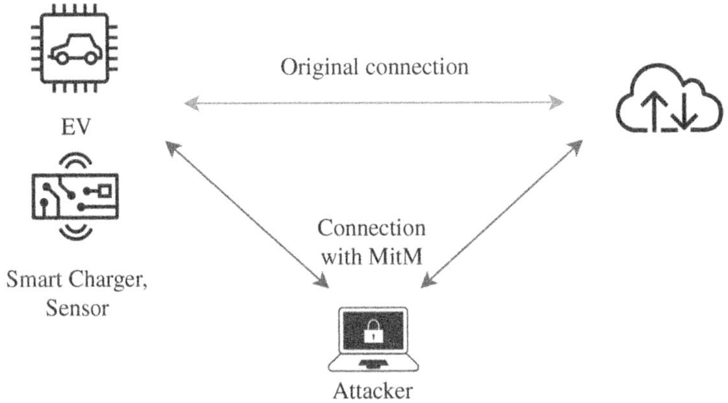

Figure 5.3 Man-in-the-middle attack.

contain vulnerabilities). This is shown by the fact that data transmission may be unencrypted and/or not require any authentication [57].

Such a system is susceptible to a MitM attack. A successful MitM attack could involve intercepting the stream of data from transformers located throughout a smart grid, replacing genuine readings with falsified readings that are below the real values, causing the total power consumption level to read below that of the actual value, which may lead to the smart charging algorithms overloading the network. Another could be simply intercepting communication from a smart charger to determine details about vehicles charging there, without modifying messages. If the attacker managed to listen to many smart chargers, it may be possible to start tracking EVs as they charge at different locations in an area. This could be performed by inserting a small plug between the EV charging cable and the charging port.

5.4.2 Attack Detection Methods

To combat the threat of attacks, detection and prevention methods need to be put into place to help thwart the possibility of a successful or harmful attack.

5.4.2.1 Abnormal State Estimation

Attacks such as data injection could put the system into a state it would not normally encounter (abnormal state). The state of the system can be observed, and it can be decided whether or not the current state is abnormal or not. If it is decided that the current state is abnormal, action can be taken to return the system to a normal state.

The authors of [58] use several Snort (an intrusion detection system [IDS]) rules to detect sensitive or abnormal operations in a smart grid network, specifically applied to smart meters, based on a number of message signatures and threat priorities. The logs are used to generate a cyber impact factor that is combined with real-time redundant measurements and used to estimate the current state of the system. A chi-square test is then applied with a given threshold to determine the presence of bad data in the current state. The IDS logs can then be used later on to track down the source.

Machine learning can be used to create a model that can be used to detect anomalous behavior in the system. In [59], the authors compare a number of feature reduction and supervised machine learning algorithms applied to false data injection, showing that the genetic algorithm (GA) for feature selection with support vector machine (SVM) classification can be combined to effectively detect attacks. Ten thousand instances of measurements were used as training, with half of them being infected as part of an FDI attack. In [60], supervised, semi-supervised, classifier and feature space fusion, and online machine learning algorithms are analyzed in multiple attack scenarios.

5.4.2.2 Message Encryption and Authentication

Smart charging will require communication of large numbers of connected devices, requiring secure messages that can be authenticated, a problem shared with the Internet of things (IoT) in general. To stop an individual intercepting and modifying a message, message encryption and authentication can be used. Message encryption is used to prevent an unauthorized individual from being able to read the message. Message authentication makes it possible to determine who sent the message and whether or not it has been tampered with while in transit.

The authors of [61] discuss various technologies such as encryption, authentication, secure firmware, and internet protocol security (IPSec) in the application of smart meters as part of an attack analysis, which could apply directly to smart chargers located throughout a smart grid. The paper proposes an authentication scheme based of the Diffie–Hellman key exchange protocol, focusing on minimal overheads for the constraint devices used in the smart grid.

While implementations of ubiquitous encryption ciphers such as advanced encryption standard (AES) can be run on constraint devices [62], there are a number of low-power encryption ciphers such as Robin, TWIS, FeW, TWINE, SIMON, and SPECK that can be read about in [63]. These protocols may be better suited to low-cost constrained devices that may be found in applications such as smart chargers or sensors as they require less hardware to implement.

5.4.2.3 Denial-of-Service Attacks

During a denial-of-service attack, attackers flood the network with packets of data to overwhelm the system (switches, routers, controllers, etc.). In a smart charging scenario, this could cause a number of problems. The attackers may be able to prevent owners from viewing pricing information [64] that can impact a user's ability to make charging decision. In [49], the authors discuss a number of ways a DoS attack can be used against a smart charging network, such as affecting EV to control center of communication and wireless radio jamming and other methods of attack.

There are ways of mitigating DoS attacks. One such method is rate limiting in which all network traffic flow is restricted. More intelligent ways to identify DoS attacks also exist. The authors of [65] employ multiple machine learning models, working together. Inputs such as source, destination, packet size, and idle time are used when analyzing the network. Shallow and deep autoencoders are used for learning features from the data that are combined using multiple kernel learning (MKL).

5.4.3 Privacy Concerns and Privacy-Preserving Methods

Privacy is a big concern in the modern age of connected devices. Sensitive personal data needs to be kept secure to stop attackers from gaining access. Many of the EV smart charging optimization algorithms depend on having access to data such as the make and model of each EV, a schedule of when it will be arriving and leaving, the SoC of the battery when it arrives and when it leaves, and even, for route planning algorithms, the intended destination and stops of a user's journey. Additionally, at public charging stations, payment will need to be taken, possibly through the charging station itself. This is all data that attackers may use to target individuals, if they gain access to it. Also, while non-smart charging is possible (for example, a simple charger plugged into a standard electrical outlet), as regulations such as those introduced earlier in this chapter are brought in, this will not be an option for the majority of EV owners.

The data required for charging can be linked back to EVs and therefore individuals. In [66], the authors showed that with access to high-resolution values, the probability of being able to link successive charge requests over multiple time slot is high, achieving an average of 97% for 140 EVs. This is due to the fact that with a high resolution of values, there are more possible values each request could take. This means that it is less likely that two requests from different vehicles will have a very similar SoC, making the probability of two requests with very similar SoCs coming from the same vehicle increase. The authors of [67] produce an EV model recognition strategy that can determine the model of an EV based on its charging characteristics.

One way to overcome this problem is described later on in [66]. Instead of each EV sending a single charge request for a given time slot containing information such as SoC and schedule data, this request is split into multiple requests for a single time slot. The data in each request will combine to form a charging priority for that time slot. Combining all of the priorities from all of the requests of an EV for a given time slot will produce the original priority, given by the actual SoC and schedule data that the EV keeps private.

Differential privacy is the technique of introducing noise to a dataset to keep the individual's data private while still providing utility with the data. The load an EV draws is obvious when observing power consumption data as it draws more power than most appliances and for longer. As bigger batteries are introduced, this affect will become greater as there will need to be higher power chargers to fill them in a reasonable amount of time. Therefore, it can be used to track a consumer's EV usage by monitoring the amount of electricity they use. Differential privacy would introduce noise into this charging data, typically by using an external battery. This battery can provide energy at random times in the day, and charge at random times in the day, introducing noise into the overall electricity consumption data, obscuring the actual usage of the consumers.

Another system is given in [66] for a distributed system for preserving privacy. A large number of EVs want to charge, and they have an aggregated node. The EVs split into groups, with each group sharing a mask. They enter their own charging request data into a table and apply the mask. All of the requests are homomorphically encrypted and are sent to the aggregating node. This node then combines the encrypted charging request data to form a total, which is then decrypted. This is broadcast to all EVs, which are then able to determine whether or not they are allowed to charge in that time slot based on the total power request for each charge request.

While it is feasible to charge an EV without any smart capability, those wanting to take part in V2G will need to allow communication from their EV/smart charger, to allow for data such as control signals and energy monitoring, and being able to identify the EV owner's identity to reimburse them or provide credit for taking part in V2G. In [68], the authors propose a system architecture for a secure smart V2G network, providing three privacy objectives: identity anonymity, vehicle untraceability, and forward privacy.

5.5 EV Smart Charging Strategies

The main goal of any EV charging strategy is to ensure that all EVs that have requested to charge meet their required charge level by the specified deadline while keeping to the limits of the grid infrastructure. If all EVs request to charge at the same time, allowing all of them to charge at once is not an option as this would overload the grid. Therefore, a charging strategy needs to be used to charge the EVs, without overloading the grid. There are various proposed strategies, each with its own advantages and limitations.

5.5.1 Optimization Approaches

There are a number of optimization approaches that can be taken, each one suiting a different environment. There are also different goals that charging station operators can target, such as minimum charge time and maximum energy efficiency (minimum energy loss).

5.5.1.1 Future Scheduling

This approach would be most appropriate for locations that will know how many EVs and their charging rate ahead of time, for instance, at home or a workplace. Customers can log into a booking system in advance of arrival, specifying their arrival time, departure time, rate of charging, etc. An optimization algorithm can then use this data to compute a charging schedule that specifies a charge start time, end time, and rate of charge, for each vehicle.

Using future scheduling, the optimum schedule can be computed, giving the best chance that all EVs are charged to their requested level by their departure time. However, this method is not particularly flexible. If EVs are coming and going, they could interrupt and even invalidate a predetermined schedule, as the new demands would have to be accounted for. This makes it unsuitable for an environment where the number of EVs charging at any one time is random, such as a public car park.

This method also brings privacy concerns. The time of arrival, location, departure time, charging power, and battery SoC are all known beforehand. This is data that may be valuable to an attacker, as it allows them to determine information such as a person's schedule, where they will be at a given time, or the maximum range from their location. For a location such as a workplace, it may be possible to estimate power demand, as historical data would provide an insight into what could be expected, removing the requirement of collecting potentially sensitive data about individuals.

5.5.1.2 Battery Health Optimization

Batteries are the highest-cost component of a modern EV. As batteries wear, their total energy capacity decreases, decreasing the range an EV can travel before recharging. This leads EV manufacturers to oversize the batteries used in EVs to account for this, increasing the cost of the EV and its weight.

An important factor not taken into account by most of the optimization methods researched so far is battery health. There are a number of factors that affect lithium-ion battery life, including temperature and SoC. Decreasing the temperature that the battery is subjected to can increase the longevity of the battery. Similarly, minimizing the amount of time a battery stays at a high SoC can also decrease wear on the battery [69].

Decreasing wear on EV batteries could offer a couple of benefits. It would be possible to decrease the size of the battery in an EV while offering the same capacity and decreasing cost and weight of the vehicle. Alternatively, it would be possible for EV manufacturers to use more of the battery capacity (not having to take into account as much wear and tear on the battery), increasing range without increasing cost or weight of the vehicle.

5.5.1.3 Energy Loss Minimization

When generating and transmitting power, some level of energy is lost through components such as cables and transformers. This is called "system loss." In distribution networks, voltage is fixed, and the value depends on the level of the network. This means as you increase power, current increases. As current increases, losses in cables increase due to the resistance of the cables. This heat increases wear on components, increasing maintenance costs. Higher losses also mean less of the electrical energy being generated is being bought by customers, increasing operation costs. The authors of [70] investigated the system losses under uncoordinated charging, finding them to be significant even under relatively low EV market penetration, reaching as high as over 3× the normal system losses with no EVs when EV penetration reached 63%. They found that by introducing their smart charging optimization algorithm, they were able to reduce the peak loads, removing the excess system losses.

5.5.2 Artificial Intelligence Approaches

The newest research area for smart charging is with the use of artificial intelligence. Using optimal historical data to train a model can generate close to optimal charging results in real time. It can also be used to help predict future grid load that can be used for charge scheduling.

5.5.2.1 Deep Learning for Smart Charging

Machine learning can be used to build predictive models that can be used to make decisions on the charging of EVs in real time. In [71], the authors presented a few models that can be used to decide whether or not to charge an EV based on data such as current price and driving, trained on optimal historical charging data.

Predicting charging loads caused by EVs can be difficult. Being able to predict immediate future loads can be useful in determining charging schedules. Knowing when there is likely to be a drop in demand would allow charging station operators to schedule more EVs in advance while avoiding the unnecessary risk of scheduling EVs and possibly not being able to fulfill the charging requests. It can also be used to predict the arrival of new EVs, allowing charging operators to accommodate accordingly. The authors of [72] proposed a method to forecast EV load using deep learning. The authors of [73] use deep reinforcement learning to determine an optimal strategy for EV charging navigation.

5.5.2.2 Predicting Charging Profiles

A problem with scheduling EV charging and making the most use out of the available power capacity is the charging profiles of different EVs. The charging rate of a large number of EVs is limited by the capacity of the grid infrastructure, so achieving the most efficient use of available capacity is important. The rate at which an EV charges depends on a number of factors, including current SoC, temperature, and voltage. The simplest charging profile would assume that an EV charges at full power until it reaches its requested SoC. However, in reality, this is not the case. For example, EVs generally taper off the charging power as they reach higher states of

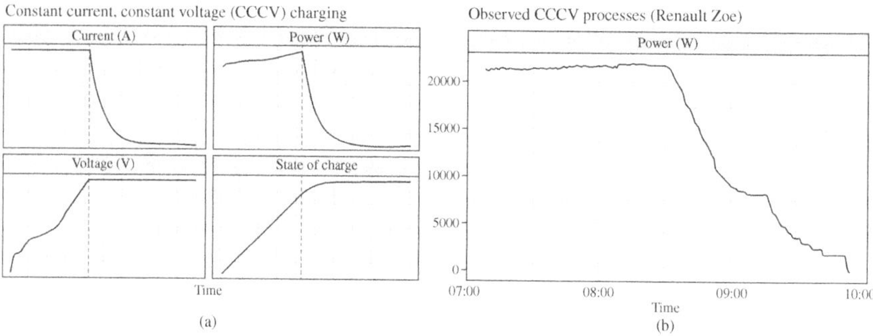

Figure 5.4 Simulated versus measured EV charge curve [74]. (a) Simulated CCCV process consisting of two distinct stages. (b) Observed CCCV process from the perspective of the charge point operator (CPO).

charge to minimize battery wear, as can be seen in Figure 5.4. Getting the most out of the available grid capacity means being able to organize charging around battery profiles to keep the total demand from all EVs as close to maximum capacity as possible.

Charging profiles for various EV batteries are generally not publicly known. In [74], the authors propose a data-driven method for predicting the charge profile of an EV battery to maximize usage of available power.

5.5.3 Coordinated Charging

For smart charging to be effective, the charging of all of the EVs on the network needs to be coordinated.

There are two approaches to this problem, centralized and distributed optimization. Examples of both will be provided in this section. In [75], a comparison is provided between the two approaches.

5.5.3.1 Centralized Optimization

One method of coordinating smart charging is through centralized charging orchestration. All EVs communicate with the centralized coordinator, supplying relevant data such as start time, end time, maximum charge power, etc. The coordinator then computes an optimal schedules for all EVs. Each EV is then sent its schedule that it follows.

In [76], a reservation-based system is used, also taking into account wait times and user discontent for that particular station. Customers make use of an app to browse available charging stations and view the number of stations available, estimated charging costs, user ratings, waiting time, etc. The user will input their requested arrival time, starting SoC, required finishing SoC, and required finish time. The system then computes all optimal solutions and returns two of them to the user for them to select one. For this to function optimally, accurate details about the users EV need to be known. As not all drivers are aware of the full specifications of their vehicle, an automated method of obtaining this data may be a worthwhile addition.

An online solution that does not depend on any future scheduling can be seen in [77]. The online algorithm is based on an offline algorithm that minimizes a cost function to compute the optimum charging rate for each EV for each time period. The online algorithm computes the optimal rate for each EV for each time period, presuming that no EVs will arrive in the future. As can be seen in Ref [34], there is a difference in performance between the offline and online algorithm; however it is minimal (no more than 6%).

5.5.3.2 Distributed Optimization

Another method of coordinated smart charging is to solve the problem with distributed optimization [78]. Smaller problems are solved throughout the network, and, through coordination, a global solution is reached that satisfies the needs of all individual EVs and of the grid. Centralized approaches require a significant amount of computation, which increases as network size increases. They also require coordination and communication from each EV to the centralized controller. As distributed solutions limit the amount of data sent, they lower the communication requirements and decrease the attack surface. Distributed systems also increase the resilience to attack as part of the network can go down without affecting the rest of the network.

An active approach is taken in [79], where each EV aggregator is concerned with meeting the requirements of the EVs at that facility. Each aggregator computes its optimal charging schedule. This is then passed to the distribution operator at the grid level, which will then compute an optimal schedule based on its own concerns, the capacity of the grid, and keeping variables within acceptable limits. The operator then sends its optimal solution to each aggregator for them to recompute their optimal charge schedule. This is repeated until some convergence criteria are met.

In [80], the authors propose a solution in which each EV communicates with its local network transformer, using day-ahead information. A restriction is placed on the charging, in that when an EV starts charging, it continues until it completes, and the power at which the EV charges is constant that can help increase the life of the EV batteries. The transformer receives the proposed charging profile from each EV, which consists of a map detailing the charge rate of that EV for each time slot. The transformer will then compute a control signal, e.g. electricity price, to all EVs. The EVs then produce a new charging profile that is sent to the transformer. This process repeats until convergence.

Another distributed approach is covered in [81], where each charger is responsible for its own charging schedule. The user inputs their requirements, such as required charge level and next departure time. The charger then decides on a schedule based on the voltage of the local distribution network. The current network voltage allows each charger to approximate the current load on the grid and the remaining capacity. To charge, the power will increase if there is enough grid capacity. As load on the grid increases, the rate at which the charger increases will decrease. If the grid reaches capacity, the chargers will back off. This approach requires no communication between any of the chargers.

5.5.4 Population-Based Approaches

Game theory allows the modeling and observation of the interactions between a number of individuals, with the goal of optimizing an objective. Mean-field game (MFG) theory allows the modeling of large populations in which each individual's actions have a negligible impact on the system. The population's mean then impacts the decision of each individual. In this way, the population is viewed as a continuum instead of individuals, meaning individuals' decision in addition to their state also depends on the population's distribution. The goal is to obtain the optimum strategy, which is found once Nash equilibrium is reached. In this situation, no individual wishes to deviate from their optimal strategy while other players follows their own optimal strategies.

MFGs are well suited to the problem of EV smart charging as we want to model large number of EVs and their effect on the grid. Individually, each EV will have a negligible impact on the grid and its operation. However, the population's average changes individual's strategy, and changing the strategy can lead to changes in population's mean and consequently in other EV strategies.

The authors of [82] used MFGs to simulate a charging station able to charge up to 500 EVs. The result was that the EVs were charged in a steady manner, keeping the load on the grid relatively steady. Some of the EVs overshot their target charge time of 120 minutes by 27 minutes, although it can be seen that most were charged to a high capacity at that point as the load on the grid drops significantly. To combat this, the cost of finishing late is increased, dropping 12 minutes from the overshoot time.

A number of other papers take different approaches to applying MFGs to EV smart charging. In [83], the authors take into consideration the availability of the EVs to supply energy back to the grid. The authors of [84] use MFGs to optimize the revenue of electric vehicle charging stations that use renewable power generation (RPG) as their main source of energy, selling excess energy to the grid and buying energy from the grid when their RPGs are not sufficient to supply the current EV demands. This chapter also takes the viewpoint of charging at centralized stations rather than at home. In [85], a cost of battery degradation based on the current the EV charges with is included in the algorithm.

A generalization made by most of the papers examined is that EVs are the same in their properties, such as battery capacities and charging rates. However, in the real world, properties that affect charging, such as total battery capacity and charging rates, vary from manufacturer to manufacturer and can even vary within a manufacturer based on the target use case of a model (such as a small urban EV versus a bigger longer range EV) and also the different trim levels available for each model. The authors of [85] examined a scenario with 10 types of plug-in electric vehicles (PEVs), showing a large difference in the strategies chosen by the different categories of PEVs.

5.5.4.1 Case Study

As part of the work on this chapter, the authors put together a case study in which an MFG using McKean–Vlasov dynamics. In this example, we present an economic analysis of integration of EVs into smart cities. Assume we have N homogeneous and competitive EV owners who decide what is the optimal battery level for their car at all time. Their decision is based on an additive utility function that depends on three parameters:

- The cost of charging or discharging batteries.
- Psychological cost of operating on the battery as a function of time.
- Disincetivizing the operation of EVs with near-empty battery level.

This notation follows the notation of [86]. We assume we have N players $\{1, 2, \dots, N\}$. The level of battery for player i at time t is represented by $X_t^{(i)}$. The change in the state variable $X_t^{(i)}$ can be represented as a difference equation:

$$X_{t+1}^{(i)} = X_t^{(i)} + \Delta(\alpha(t) - g(t)), \tag{5.1}$$

where g is the amount of electricity consumption for the EV and α is the decision variable for the user (the amount of charge or discharge). We restrict both α and g such that the battery level stays between 0 and 1. We assume each player has access to the distribution of battery level m_t^N for all cars at time t:

$$m^N(t, X) = \frac{1}{N}\sum_{i=1}^{N} \delta_{X_t^{(i)}} \tag{5.2}$$

The objective for each EV is to solve an optimization problem subject following the dynamics in (5.1)

$$v(0, X) = \sup_{\alpha_t} \mathbb{E}\left(\sum_{i=1}^{N} -C\left(X_t, \alpha_t, m_t, t\right)\right), \tag{5.3}$$

where υ represents the value function. We explicitly define C as summation of following terms:

$$C(\alpha_t, X_t, m_t, t) = \alpha_t p_t(m_t) + h(t, \alpha_t) + f(t, X_t), \tag{5.4}$$

where $f(t, X_t) = w_1(1 - X_t)^4$ incentivizes the owners to avoid emptying their battery and $h(t, \alpha_t) = \frac{1}{2} w_2 \alpha^2$ shows the unwillingness of the car owner to buy or sell energy at time t. The price p_t is considered as a function of expected consumption:

$$p_t = \left(\left[g_t + \frac{d}{dt} \int X_m(t, dX) \right]^+ + d_t \right)^2 \tag{5.5}$$

where g_t and d_t are mean energy consumption and total energy consumption as a function of time that is presented in Figure 5.5 [86].

The notion of Nash equilibrium means a set of strategies where none of the players has an incentive to unilaterally deviate from their optimal strategy. In this work, we use mean-field equilibrium as an approximation of the Nash equilibrium. The mean-field equilibrium for a player given the mean-field distribution of the agents is a set of strategies that no player has incentive of deviating from their optimal strategy. It is worth noting that, in this scenario, as we assumed the players are homogeneous, the strategy is symmetrical across all players. For solving this optimization, we restrict ourselves to Markovian strategies.

To solve the optimization, we follow the backward recursive algorithm presented in [87]. In this way, we initialize the value function $\upsilon_{T+1}(m_{T+1}, X_{T+1}) = 0$ and then go backward in time. For all possible mean-field distributions, we solve the following fixed point equation:

$$\tilde{\gamma}(.|x^{(i)}) \in \underset{\gamma(.|x^{(i)})}{\operatorname{argmax}} \, \mathbb{E}^{\gamma(.|x^{(i)})} \left[-C\left(x^{(i)}, A^i, m^N\right) + \delta V\left(\phi\left(m^N, \tilde{\gamma}\right), X^{i'}\right) | m^N, x^{(i)} \right] \tag{5.6}$$

$$V\left(m^N, x^{(i)}\right) = \mathbb{E}^{\tilde{\gamma}_t(.|x^{(i)})} \left[-C\left(x^{(i)}, A^i, m^N\right) + \delta V\left(\phi\left(m^N, \gamma_t\right), X^{(i')}\right) | m^N, x^{(i)} \right] \tag{5.7}$$

where γ_t represents the strategy of each agent. For a given prescription function γ_t, the statistical mean field m_t^N evolves according to the discrete time that can represented by function ϕ where $m_{t+1}^N = \phi\left(m_t^N, \gamma_t\right)$. The strategy structured from following the algorithm is a mean-field equilibrium for the game. In solving the dynamic programming problem, we consider a finite state system with five states as an abstraction for the dynamic system represented in (5.1). For mean-field distribution $m^N(t, X)$, we assumed each state's density can be from the set $\{0, 0.2, 0.4, 0.6, 0.8, 1\}$ where the summation of densities in all states adds up to 1. The input is selected from the set $\{-2.5, -2, \ldots, 1\}$ given the battery level always stays between 0 and 1.

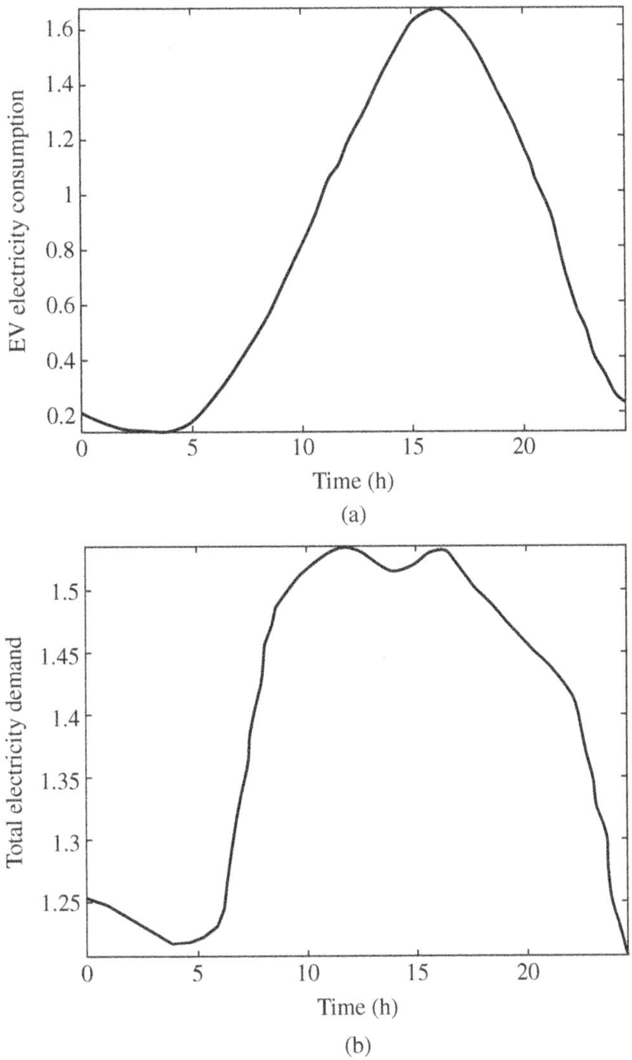

Figure 5.5 Mean energy consumption of EVs g_t (a) and total nominal electricity demand d_t(b) as a function of time for a day.

We solved the dynamic programming problem for a day. After solving the problem, we simulate the resultant strategy for 10 000 EVs. Figure 5.6 shows the density solution for the mean-field distribution, and Figure 5.7 displays the battery level of EVs. One can see the battery level for most cars remains near half full when the demand is not high. However, when the demand rises, the EVs keep their battery level below 50% to avoid paying a higher price according to Figure 5.8.

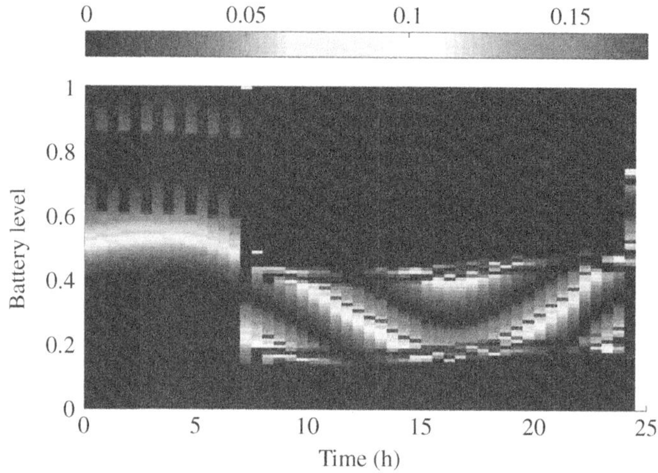

Figure 5.6 Density distribution $m^N(t, X)$ as a function of time for a day.

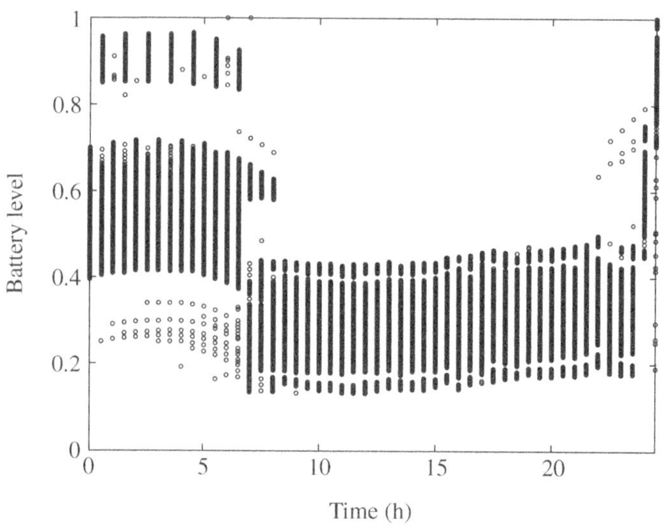

Figure 5.7 Battery level of EVs $X^i(t)(i = 1, 2, \ldots, N)$ as a function of time for a day.

5.6 Conclusion

Considering the rapidly growing uptake of electric vehicles all around the world, and by knowing that most of these new assets connect to the low-voltage electricity networks of the smart cities and communities, there is a strong need to have a

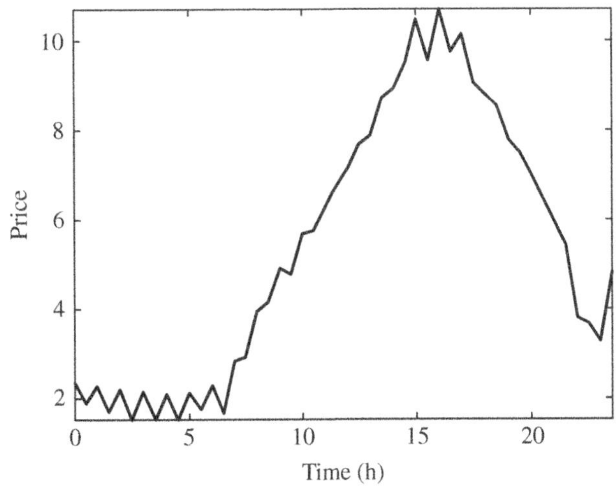

Figure 5.8 Evolution of electricity price as a function of time for a day.

systematic approach that coordinates these resources and prevents unnecessary investments in the infrastructure. Smart charging is a necessity in such conditions, increasing the flexibility of existing infrastructure, and is able to reduce investments while keeping costs low to consumers and meeting customer needs. This will also cause a well-distributed demand of electric fleet vehicles (EFVs) from the transmission point of view that also avoids new investments there.

Moreover, not only these assets are able to participate in the provision of different market-based or incentive-based network services, but also they are distributed energy sources that can be a backbone of the energy networks of a smart city in times of a wide-area failure or disaster.

The following directions need to be investigated more to pave the way for the integration of the EFVs in smart cities and to cope with the challenges in this area:

- Design of a multi-agent framework for modeling and simulation of the EV integration into smart cities is a promising direction in this field that needs to be more investigated.
- Considering the distributed nature of EVs in smart cities and their uncertain behavior, the application of artificial intelligence and machine learning techniques would be a promising direction for making data-driven decision-making platforms for the control and management of EFVs.
- Another direction is to develop a suitable market for successful integration of the EFVs into electricity grids that incentivizes their participation in the provision of different services for the power grids while also providing a better management strategy.

- The requirement for secure low-latency communication in smart charging CPSs provides another direction, as the continually increasing market share of EVs will provide bigger targets for potential attackers.

Acknowledgments

The authors are grateful to Luca Arnaboldi (University of Edinburgh) for helpful discussions and feedback on the security aspects of this work.

References

1 BBC East Midlands (2020). Wireless taxi charging to be trialled at Nottingham station. *BBC News* (January 2020).
2 Nick Stat (2019). Norway will install the world's first wireless electric car charging stations for Oslo taxis. *The Verge* (March 2019).
3 Sytner Group Limited (2019). BMW wireless charging.
4 SAE International (2020). Wireless power transfer for light-duty plug-in/electric vehicles and alignment methodology. Standard, SAE International, October 2020.
5 SAE International (2020). Communication for wireless power transfer between light-duty plug-in electric vehicles and wireless EV charging stations. Standard, SAE International, September 2020.
6 Fronzek, T. (2015) e-smartConnect: Volkswagen is conducting research on an automated quick-charging system for the next generation of electric vehicles. Received from https://www.volkswagen-newsroom.com/en/press-releases/e-smartconnect-volkswagen-is-conducting-research-on-an-automated-quick-charging-system-for-the-next-generation-of-electric-vehicles-1512.
7 United States Environmental Protection Agency (2014). Global greenhouse gas emissions data.
8 Muntean, M., Guizzardi, D., Schaaf, E. et al. (2019). Edgar booklet 2019, 08 2019.
9 Bastin, Z., Bhattacharya, S., and Kumar, A. (2019). International electric-vehicle consumer survey.
10 UK Government: legislation.gov.uk (2019). The Climate Change Act 2008 (2050 Target Amendment) Order 2019.
11 French Government: gouvernement.fr (2019). Minutes of the Council of Ministers of April 30, 2019.
12 Energy Danish Ministry of Climate and Utilities (2019). Aftale om klimalov.

13 Legislative Committee of the Parliament (Hungary) (2020). Summary statement of the argument (t/7021.) attached to the proposed amendment aimed at announcing the climate[change] crisis, which closed the committee's detailed debate. (Összegz? jelentés a klímavészhelyzet kihirdetésér?l szóló törvényjavaslathoz (t/7021.) benyújtott, részletes vitát lezáró bizottsági módosító javaslat vitájáról), May 2020.

14 Parliamentary Counsel Office (New Zealand) (2019). Climate Change Response (Zero Carbon) Amendment Act 2019.

15 Government Offices of Sweden (2017). Riksdag passes historic climate policy framework.

16 U.S. Department of Energy (2013). Qualified Plug-In Electric Vehicle (PEV) Tax Credit, Public Law 112-240.

17 Gardner, G. (2019). Tesla, GM Lose Bid to Raise Ceiling for Federal EV Tax Credit. Received from https://www.forbes.com/sites/greggardner/2019/12/17/tesla-gm-lose-bid-to-raise-ceiling-for-federal-ev-tax-credit/.

18 U.S. Department of Energy (2019). Alternative Fuel Infrastructure Tax Credit.

19 European Automobile Manufacturers Association (2020). EU and UK - Electric Vehicles: Tax Benefits and Purchase Incentives, July 2020.

20 IEA (2020). Global EV Outlook 2020. *Technical report*. IEA: Paris. https://www.iea.org/reports/global-ev-outlook-2020.

21 IEA (2020). Electric Car Stock by Region and Technology, 2013–2019.

22 HM Government (HMG) (2018). The road to zero: next steps towards cleaner road transport and delivering our industrial strategy.

23 UK Government: gov.uk (2020). Government takes historic step towards net-zero with end of sale of new petrol and diesel cars by 2030, November 2020.

24 UK Government: legislation.gov.uk. (2018). Automated and Electric Vehicles Act 2018.

25 Aghapour, R., Sepasian, M.S., Arasteh, H. et al. (2020). Probabilistic planning of electric vehicles charging stations in an integrated electricity-transport system. *Electric Power Systems Research* 189: 106698.

26 Department for Transport (2020). Electric vehicle charging device statistics: April 2020 - GOV.UK. https://www.gov.uk/government/statistics/electric-vehicle-charging-device-statistics-april-2020.

27 Admin (Zap-Map). Zap-Map EV Charging Stats 2020.

28 Kamgarpour, M., Ellen, C., Soudjani, S.E.Z. et al. (2013). Modeling options for demand side participation of thermostatically controlled loads. *2013 IREP Symposium Bulk Power System Dynamics and Control - IX Optimization, Security and Control of the Emerging Power Grid*, pp. 1–15.

29 Soudjani, S.E.Z. and Abate, A. (2015). Aggregation and control of populations of thermostatically controlled loads by formal abstractions. *IEEE Transactions on Control Systems Technology* 23 (3): 975–990.

30 Soudjani, S.E.Z. and Abate, A. (2013). Aggregation of thermostatically controlled loads by formal abstractions. *2013 European Control Conference (ECC)*, pp. 4232–4237.

31 Wooding, B., Vahidinasab, V., Kazemi, M., and Soudjani, S. (2021). Control and management of active buildings. In: *Active Building Energy Systems: Operation and Control* (eds. V. Vahidinasab, Behnam Mohammadi-Ivatloo), 1. Springer.

32 Wooding, B., Vahidinasab, V., and Soudjani, S. (2021). Operation and control of a population of active buildings at network level. In: *Active Building Energy Systems: Operation and Control* (eds. V. Vahidinasab, Behnam Mohammadi-Ivatloo), 1. Springer.

33 Wang, M., Shen, X.S., and Zhang, R. (2016). *Mobile Electric Vehicles*. Springer.

34 Giorgio, A.D. and Liberati, F. (2015). A model predictive control-based approach for plug-in electric vehicles charging: power tracking, renewable energy sources integration and driver preferences satisfaction. In: *Plug in Electric Vehicles in Smart Grids* (eds. S. Rajakaruna, F. Shahnia, A. Ghosh), 203–240. Springer.

35 Falk, R. and Fries, S. (2012). Electric vehicle charging infrastructure security considerations and approaches. *Proceedings of INTERNET*, pp. 58–64.

36 Wang, Y.-n., Lin, Z., Liang, X. et al. (2016). On modeling of electrical cyber-physical systems considering cyber security. *Frontiers of Information Technology and Electronic Engineering* 17: 465–478.

37 Kim, Y., Kolesnikov, V., and Thottan, M. (2013). TSAF: Tamper-resistant and scalable mutual authentication framework for plug-in EV charging. *2013 IEEE International Conference on Smart Grid Communications (SmartGridComm)*, pp. 444–449.

38 Nayak, A., Rana, R., and Mishra, S. (2019). Frequency regulation by electric vehicle during grid restoration using adaptive optimal control. *IFAC-PapersOnLine* 52 (4): 270–275. *IFAC Workshop on Control of Smart Grid and Renewable Energy Systems CSGRES 2019*.

39 Wooding, B., Vahidinasab, V., and Soudjani, S. (2020). Formal controller synthesis for frequency regulation utilising electric vehicles. *2020 International Conference on Smart Energy Systems and Technologies (SEST)*, pp. 1–6.

40 Habibi, M., Vahidinasab, V., and Sepasian, M.S. (2020). Application of mobile energy storage to facilitate energy transfer between TSO and DSO networks. *2020 10th Smart Grid Conference (SGC)*, pp. 1–3.

41 Finn, P., Fitzpatrick, C., and Connolly, D. (2012). Demand side management of electric car charging: benefits for consumer and grid. *Energy* 42 (1): 358–363. *8th World Energy System Conference, WESC 2010*.

42 Huang, Q., Jia, Q., Xia, L. et al. (2014). EV charging load scheduling following uncertain renewable energy supply by stochastic matching. *2014 IEEE International Conference on Automation Science and Engineering (CASE)*, pp. 137–142.

43 Huang, Q., Jia, Q., Qiu, Z. et al. (2015). Matching EV charging load with uncertain wind: a simulation-based policy improvement approach. *IEEE Transactions on Smart Grid* 6 (3): 1425–1433.

44 Hu, J., Morais, H., Sousa, T., and Lind, M. (2016). Electric vehicle fleet management in smart grids: a review of services, optimization and control aspects. *Renewable and Sustainable Energy Reviews* 56: 1207–1226.

45 NewMotion (Shell Group). The Electric Range of an EV.

46 Energy.gov (2016). Fact 939: August 22, 2016 all-electric vehicle ranges can exceed those of some gasoline vehicles, August 2016.

47 Liu, H., Yin, W., Yuan, X., and Niu, M. (2018). Reserving charging decision-making model and route plan for electric vehicles considering information of traffic and charging station. *Sustainability* 10 (5): 1324.

48 Peng, C., Zou, J., and Lian, L. (2017). Dispatching strategies of electric vehicles participating in frequency regulation on power grid: a review. *Renewable and Sustainable Energy Reviews* 68: 147–152.

49 Antoun, J., Kabir, M.E., Moussa, B. et al. (2020). A detailed security assessment of the EV charging ecosystem. *IEEE Network* 34 (3): 200–207.

50 David, M. (2011). Hacking the lights out. *Scientific American - SCI AMER* 305: 70–75.

51 Arnaboldi, L., Czekster, R.M., Morisset, C., and Metere, R. (2020). Modelling load-changing attacks in cyber-physical systems. *Electronic Notes in Theoretical Computer Science* 353: 39–60.

52 Humayed, A., Lin, J., Li, F., and Luo, B. (2017). Cyber-physical systems security–a survey. *IEEE Internet of Things Journal* 4 (6): 1802–1831.

53 Mazur, M. and Kurkowski, M. (2019). Complexity of security protocols verification tools. *2019 IEEE 15th International Scientific Conference on Informatics*, pp. 000403–000408.

54 Bhusal, N., Gautam, M., and Benidris, M. (2020). Cybersecurity of electric vehicle smart charging management systems. *2020 52nd North American Power Symposium (NAPS)*, pp. 1–6. IEEE

55 Porter, J. (2020). Amazon says it mitigated the largest DDoS attack ever recorded, June 2020.

56 Newman, L.H. (2018). A 1.3-Tbs DDoS Hit GitHub, the Largest Yet Recorded, March 2018.

57 Babu, B., Ijyas, T., Muneer, P., and Varghese, J. (2017). Security issues in SCADA based industrial control systems. *2017 2nd International Conference on Anti-Cyber Crimes (ICACC)*, pp. 47–51.

58 Liu, T., Sun, Y., Liu, Y. et al. (2015). Abnormal traffic-indexed state estimation: a cyber–physical fusion approach for smart grid attack detection. *Future Generation Computer Systems* 49: 94–103.

59 Sakhnini, J., Karimipour, H., and Dehghantanha, A. (2019). Smart grid cyber attacks detection using supervised learning and heuristic feature selection. *2019 IEEE 7th International Conference on Smart Energy Grid Engineering (SEGE)*, pp. 108–112.

60 Ozay, M., Esnaola, I., Vural, F.T.Y. et al. (2016). Machine learning methods for attack detection in the smart grid. *IEEE Transactions on Neural Networks and Learning Systems* 27 (8): 1773–1786.

61 Fouda, M.M., Fadlullah, Z.M., Kato, N. et al. (2011). A lightweight message authentication scheme for smart grid communications. *IEEE Transactions on Smart Grid* 2 (4): 675–685.

62 Bos, J.W., Osvik, D.A., and Stefan, D. (2009). Fast implementations of AES on various platforms. *IACR Cryptology ePrint Archive* 2009: 501.

63 Hatzivasilis, G., Fysarakis, K., Papaefstathiou, I., and Manifavas, H. (2018). A review of lightweight block ciphers. *Journal of Cryptographic Engineering* 8: 1–44.

64 Li, Y., Wang, R., Wang, P. et al. (2012). Resilient PHEV charging policies under price information attacks. *2012 IEEE 3rd International Conference on Smart Grid Communications (SmartGridComm)*, pp. 389–394.

65 Ali, S. and Li, Y. (2019). Learning multilevel auto-encoders for DDoS attack detection in smart grid network. *IEEE Access* 7: 108647–108659.

66 Baza, M., Pazos-Revilla, M., Nabil, M. et al. (2020). Privacy-preserving and collusion-resistant charging coordination schemes for smart grid.

67 Dang, Q., Huo, Y., and Sun, C. (2018). Privacy preservation needed for smart meter system: a methodology to recognize electric vehicle (EV) models. In *2018 IEEE Innovative Smart Grid Technologies - Asia (ISGT Asia)*, pp. 1016–1020.

68 Saxena, N., Grijalva, S., Chukwuka, V., and Vasilakos, A.V. (2017). Network security and privacy challenges in smart vehicle-to-grid. *IEEE Wireless Communications* 24 (4): 88–98.

69 Smith, K., Markel, T., Kim, G.-H., and Pesaran, A. (2010). Design of electric drive vehicle batteries for long life and low cost. *IEE Workshop on Accelerated Stress Testing and Reliability (ASTR)*, pp. 6–8.

70 Deilami, S., Masoum, A.S., Moses, P.S., and Masoum, M.A.S. (2011). Real-time coordination of plug-in electric vehicle charging in smart grids to minimize power losses and improve voltage profile. *IEEE Transactions on Smart Grid* 2 (3): 456–467.

71 López, K.L., Gagné, C., and Gardner, M. (2019). Demand-side management using deep learning for smart charging of electric vehicles. *IEEE Transactions on Smart Grid* 10 (3): 2683–2691.

72 Zhang, X., Chan, K.W., Li, H. et al. (2020). Deep-learning-based probabilistic forecasting of electric vehicle charging load with a novel queuing model. *IEEE Transactions on Cybernetics* 51 (6): 1–14.

73 Qian, T., Shao, C., Wang, X., and Shahidehpour, M. (2020). Deep reinforcement learning for EV charging navigation by coordinating smart grid and intelligent transportation system. *IEEE Transactions on Smart Grid* 11 (2): 1714–1723.

74 Frendo, O., Graf, J., Gaertner, N., and Stuckenschmidt, H. (2020). Data-driven smart charging for heterogeneous electric vehicle fleets. *Energy and AI* 1: 100007.

75 Cheng, A.J., Tarroja, B., Shaffer, B., and Samuelsen, S. (2018). Comparing the emissions benefits of centralized vs. decentralized electric vehicle smart charging approaches: a case study of the year 2030 California electric grid. *Journal of Power Sources* 401: 175–185.

76 Vaidya, B. and Mouftah, H.T. (2020). Smart electric vehicle charging management for smart cities. *IET Smart Cities* 2 (1): 4–13.

77 Guo, L., Erliksson, K.F., and Low, S.H. (2017). Optimal online adaptive electric vehicle charging. *2017 IEEE Power Energy Society General Meeting*, pp. 1–5.

78 Vahidinasab, V. (2014). Optimal distributed energy resources planning in a competitive electricity market: multiobjective optimization and probabilistic design. *Renewable Energy* 66: 354–363.

79 Fahmy, S., Gupta, R., and Paolone, M. (2020). Grid-aware distributed control of electric vehicle charging stations in active distribution grids. *Electric Power Systems Research* 189: 106697.

80 Gan, L., Topcu, U., and Low, S.H. (2012). Stochastic distributed protocol for electric vehicle charging with discrete charging rate. *2012 IEEE Power and Energy Society General Meeting*, pp. 1–8.

81 Xia, L., Mareels, I., Alpcan, T. et al. (2014). A distributed electric vehicle charging management algorithm using only local measurements. *ISGT 2014*, pp. 1–5.

82 Zhu, Z., Lambotharan, S., Chin, W.H., and Fan, Z. (2016). A mean field game theoretic approach to electric vehicles charging. *IEEE Access* 4: 3501–3510.

83 Couillet, R., Perlaza, S.M., Tembine, H., and Debbah, M. (2012). Electrical vehicles in the smart grid: a mean field game analysis. *IEEE Journal on Selected Areas in Communications* 30 (6): 1086–1096.

84 Lee, W., Xiang, L., Schober, R., and Wong, V.W.S. (2015). Electric vehicle charging stations with renewable power generators: a game theoretical analysis. *IEEE Transactions on Smart Grid* 6 (2): 608–617.

85 Tajeddini, M.A. and Kebriaei, H. (2019). A mean-field game method for decentralized charging coordination of a large population of plug-in electric vehicles. *IEEE Systems Journal* 13 (1): 854–863.
86 Couillet, R., Perlaza, S.M., Tembine, H., and Debbah, M. (2012). A mean field game analysis of electric vehicles in the smart grid. *2012 Proceedings IEEE INFOCOM Workshops*, IEEE, pp. 79–84.
87 Vasal, D. (2019). Markov perfect equilibria in non-stationary mean-field games. *Available at SSRN 3385602.*

6

Risk-Aware Cyber-Physical Control for Resilient Smart Cities

Eman Hammad[1] *and Abdallah Farraj*[2]

[1] *Department of Computer Science and Information Systems, Texas A&M University-Commerce/RELLIS Campus, Bryan, TX, USA*
[2] *Department of Electrical Engineering, Texas A&M University-Texarkana/RELLIS Campus, Bryan, TX, USA*

6.1 Introduction

Smart cities integrate sensory, actuation, and intelligent system to achieve human-centric socioeconomic goals. With multiple systems being integrated, we abstract a smart city as a system of cyber-physical systems (CPS). With a "System of Systems" approach, we focus on the study of the CPS resilience in a smart city. Resilience describes the ability of the CPS to withstand and quickly recover from adverse conditions. Adverse condition could manifest as cyber and/or physical incidents that could lead to operational disruptions of a CPS. Improving the ability of critical CPS to recover quickly and efficiently from such incidents leads to a flourishing and more reliable smart city ecosystem.

The continuously evolving nature of advanced and complex threats targeting CPS systems is challenging existing approaches to cybersecurity detection, response, and recovery. This is motivating the investigation of new and innovative cyber resilience approaches at different levels of emerging CPS within a smart city such as the smart grid. The complexity of interconnected CPS systems in a smart city offers a multitude of challenges and opportunities when addressing issues of CPS security and resilience. While CPS vulnerabilities are as-of-yet not fully understood, the ability to effectively coordinate both cyber and physical entities offers opportunities to reinforce the CPS capacity against unknown and complex threats. Developing new and innovative approaches to CPS security and resilience should consider realistic systems and then expand the applicability of that approach to other CPS systems in the smart city. In this study, we focus on smart grids due to its criticality in enabling other CPS systems and also

Cyberphysical Smart Cities Infrastructures: Optimal Operation and Intelligent Decision Making, First Edition. Edited by M. Hadi Amini and Miadreza Shafie-khah.

because of its distinguishing characteristics. Operational reliability requirements of smart grids limit the ability to apply traditional approaches to cybersecurity, specifically when considering time-sensitive processes. It is important here to note that this approach is applicable to other CPS systems in a smart city, with adjustments, according the specific CPS dynamics and capacity. Furthermore, the software-defined network (SDN) part of the proposed approach will become increasingly relevant with increased adoption of 5G networks that heavily rely on SDN for its core operation. 5G deployments continue to be rolled out and will increasingly be used to support more of the critical infrastructures in a smart city.

Various research thrusts recently investigated the quality of service (QoS) of communication systems specifically when communication links are part of system operation and control [3–6]; delay-sensitive controllers that require best effort or guarantees on meeting QoS requirements are investigated, and the communication routing problem in a smart grid is formulated as a QoS routing problem. QoS routing problem is investigated in [3] for demand-response applications, a QoS metric is developed based on pricing-based control constraints, and a greedy algorithm is proposed to find the shortest path under the QoS constraints. Smart grid QoS routing is investigated in [4], and an enhanced genetic algorithm with ticket-based flooding discovery is proposed. The works of [7–9] investigate using genetic algorithms, flocking based routing, and potential games to mitigate denial-of-service (DoS) attacks on communication links.

We extend our previous work in [10, 11], where we investigated benefits of adaptive routing on the cyber layer that considers both cyber risks and delay requirements (QoS) of cyber-physical control. We consider a threat model where an adversary targets cyber (communication) links through initiating malicious faults and jeopardizing transmitted data and/or physical systems to negatively affect system stability. We propose a cross-layer risk-aware cyber-physical control to improve the systems' resilience against cyber and physical disruptions. We combine adaptive QoS software-defined networks (SDN) routing on the cyber layer with adaptive distributed control [12, 13] on the physical layer in a new approach that strives for risk avoidance on the cyber level while satisfying physical system delay requirements. The cross-layer approach extends to the adaptive distributed control where resulting delay in transmitted measurements and commands will be utilized to regulate the control output to maintain system stability.

Our mitigation approach capitalizes on recent and accelerating developments in SDN technologies in which a communication network's data plane is decoupled from a system-aware control layer, which enables an optimal routing capability in the SDN control plane. Thus, we propose an algorithmic approach to solve the double constrained shortest path (CSP) problem that is used to formulate smart city routing while mitigating communications' link insecurities and maintaining QoS constraints.

The previous work for QoS-guaranteed end-to-end delay for multimedia delivery using SDN framework is presented in [14], where a combination of dynamic routing, scheduling, and pipelining is used. The work in [15] shows that controlling packet schedulers results in enhanced quality of experience. Dijkstra algorithm is used in network routing protocols [16, 17] to find the shortest routing paths between communication nodes. Each communication link has an associated cost related to its status value in this algorithm, and the shortest path between two nodes is calculated using the link costs. Practical implementations of Dijkstra algorithm assign link costs in a naive manner because of the difficulty in obtaining link status updates (and thus costs) for the entire communication network automatically and dynamically. As an example, consider the Open Shortest Path First (OSPF) routing where the link costs are defined as the reference bandwidth divided by interface bandwidth, and this leads to the static cost values [17]. Link costs are defined in other examples as 1, which reduces the shortest path weights to simply the hop count.

SDN technology formulation provides a methodology to solve for the shortest communication path problem using dynamic link status updates through the overall network monitoring capability in the SDN. Thus, the implementation of double-constrained QoS routing that maintains a situational awareness of the dynamic communication links' risk state within the communication network is possible using the SDN framework. Risk levels of communication links can be evaluated using specific communication link attributes and also historical data obtained from intrusion detection systems or cyber monitoring tools. Communication link attributes (such as historical data, delay, bandwidth, ownership, link type, etc.) can be collected by the SDN routing controller to be used in the routing algorithm. Such link information enables more efficient, safe, and reliable path selection.

6.2 System Model

Consider a smart power grid system as a *generic* CPS in a smart city. Let the number of buses (or *nodes*) in the smart grid be termed N. Also, consider the existence of a communication network in a topology that parallels the power grid in the smart city and connects the N buses as shown in Figure 6.1. Consider a weighted and undirected graph $G(V, E, w)$ that describes the communication network as an M-link and N-node network. In this notation, $V = \{v_1, \dots, v_N\}$ refers to the node set in G and denotes the buses of the network, $E = \{e_{ij}, i, j = 1, 2, \dots, M\}$ is the edge set of G and denotes the communication links, and w is the weight vector of G. Also, it consists of the *costs* assigned to the communication links in the net-

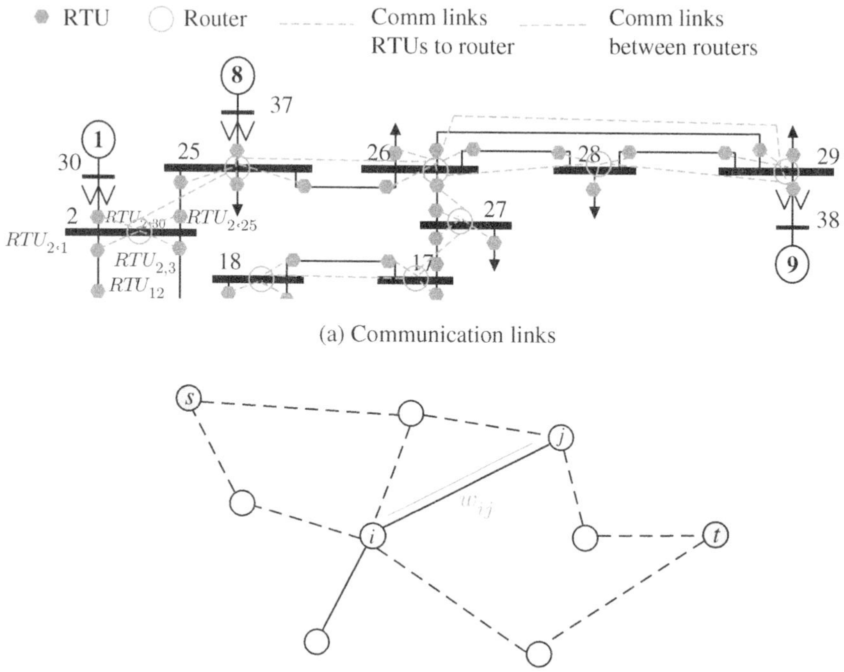

(a) Communication links

(b) Communication graph

Figure 6.1 (a) A depiction of the communication network topology for part of the 39-bus grid system and (b) a communication network routing graph highlighting link weight.

work. Furthermore, let $i, j \in E$, and then the adjacency matrix A of $G(V, E, w)$ is defined as

$$A_{i,j} = \begin{cases} 0 & i = j \\ w_{ij} & i \neq j \end{cases} \tag{6.1}$$

Let s be a source node and t be a destination node in V. We consider the routing problem between s and t in graph $G(V, E, w)$. Dijkstra algorithm is well-known for its simplicity and optimality in finding shortest path solutions between a pair of points. This algorithm will be used a benchmark in our analysis and numerical analysis.

6.2.1 Communication Latency in Smart Grid Systems

Because of the critical nature of cyber-enabled control functions within smart power grids, such CPS typically require information delivery within narrow delay requirements. Thus, the problem of finding routing paths that meet specific communication latency constraints can be investigated using CSP formulations. Furthermore, when the problem formulations includes two or more constraints, the CSP problem can be referred to as a multi-constrained shortest path (MCSP) problem that are mostly solved using algorithmic approaches to find feasible solutions [18].

Within the CSP-type framework, it is essential to characterize constraints and their interdependencies to determine whether they can be combined in some way. To illustrate, an interdependence can be clearly observed between link congestion and packet drop constraints; in addition, a similar pattern can be deduced between constraints on link congestion and communication latency. How or whether to combine constraints depends on the specific problem considered.

The use of networked sensors and control functions in smart grid systems leads to the unique constraints on data communications. In this study, we focus on the interplay between communication latency in a smart grid and CPS resilience. It is known that enhanced security could hamper network (delay) performance in traditional networks. We believe that the SDN framework is appropriate for smart grid applications because it promotes distinct approaches to system resilience and security due to the improved situational awareness and decoupled network control.

An SDN framework facilitates situational awareness through collecting performance metrics that are relevant to the communication links in smart grids. Let A_d denote a *delay cost matrix*; we define a dynamic delay cost matrix as $A_d^i = A_d(T_{sd} \cdot i)$ that is sampled from network elements at predefined intervals with a sampling period T_{sd}. Similarly, a corresponding risk cost matrix A_r can be constructed given a risk metric. Hence, an MCSP formulation can be used to model the smart grid's QoS routing problem as detailed below.

6.2.2 Risk Model for Communication Links

The attack surface of the smart grid is increased due to the integration of cyber-connected sensory and control technologies. The confidentiality–integrity–availability (CIA) framework presents a concise vulnerability classification, which can be utilized for threat modeling and the construction of suitable metrics to enable risk-aware adaptive operation. We present a risk metric to capture residual risks of in communication links taking into account vulnerabilities, applied security controls, and cyber incident history, and this will enable us to develop a corresponding network response to threats and cyber attacks targeting certain communication links. It is to be emphasized that risk is not a binary characteristic of communication links. Thus, simple qualitative labels of a link as a low risk or a high risk may not be very informative metrics to be used in system operation; hence, a more quantifiable metric would be more useful.

Risk metrics generally aim to capture the impact of an event and the probability that such an event would occur [19]. Following the same rational for communication networks in a CPS of smart city's network infrastructure, we consider a view to combine some of the network's link attributes, existing security controls and history to develop a relevant risk metrics. Considering the network link between nodes i and j, the following attributes of the link are of particular interest and represent salient qualities to consider for secure routing:

(1) Link history L_H^{ij}: If the link between nodes i and j was previously targeted by an attacker, this link would be considered more likely to be targeted again by an adversary or a similar attacks resulting in a higher risk.
(2) Installed security measures/controls L_S^{ij}: If the link between nodes i and j has strong encryption controls, this link would be considered less vulnerable to certain classes of attacks, resulting in a lower probability of attack and hence lower risk value.
(3) Inherent risks L_M^{ij}: Operational properties of the system such as physical-channel modality or channel ownership and control are represented in this model as attributes that carry a certain level of risk. For example, a communication medium that is more accessible to intrusion would be carry a higher inherent risk as, in the case of wireless communication links, vulnerability to intentional electromagnetic interference threats.

Although some interdependencies exist among the attributes, we neglect them for practical simplicity. As will be discussed later, we model L_H^{ij} as a Markov process. The link attribute L_S^{ij} quantifies the strength of the associated link security technology. We therefore consider mappings that monotonically increase with decreased security strength. As an example, let $\{Encr_1, Encr_2, Encr_3\}$ denote three possible encryption mechanisms for the communication link between nodes i and j, and consider the case of $Encr_1 \rhd Encr_2 \rhd Encr_3$, where $\rhd$ is a comparison operator with the security level of the right side's mechanism being weaker than the left one. Thus, a possible assignment for L_S^{ij} involves a ranked value of security mechanism strength that yields $L_S^{ij} \in \{1, 2, 3\}$.

In a similar fashion, when considering the inherit risk and the example of a physical channel's modality, a wireless communication channels could be deemed more vulnerable medium than Ethernet channels that are considered more vulnerable than fiber optics when considering certain threat models. The based on the predicted effort to apply a relative attack, a ranked metric may be employed. Hence, for the following channel mediums (fiber optics, Ethernet, wireless), $L_M^{ij} \in \{1, 2, 3\}$.

To combine these communication link attributes into a single risk metric L_R^{ij}, we aim to keep the general monotonic relationship between metric value and risk level. We therefore base the risk metric definition on the CIA framework.

Let C_A, C_I, C_C be defined as the *cost on availability*, *cost on integrity*, and *cost o confidentiality*, respectively, for a communication link. Then, the risk metric L_R^{ij} is proposed as follows:

$$L_R^{ij} = L_H^{ij} \times (C_A^{ij} + C_I^{ij} + C_C^{ij}) \tag{6.2}$$

Let the individual costs C_A, C_I be functions of relevant attributes of the communication link such that $C_A^{ij} = \alpha L_S^{ij}$ and $C_I^{ij} = \beta L_M^{ij}$, where α and β are appropriate weights that scale each attribute to reflect the impact of the corresponding threat. Therefore:

$$L_R^{ij} = L_H^{ij} \times (\alpha L_S^{ij} + \beta L_M^{ij}) \tag{6.3}$$

If the threat model of the communication link is focused on DoS attacks, then L_M^{ij} is of particular interest in this risk metric with the corresponding weight larger. In addition, in the case of homogeneous communication networks with similar values of L_S^{ij} and L_M^{ij} for each network link i, j, the attack history of a communication link would be the dominant term in L_R^{ij}. Furthermore, the dependencies existing among the link attributes should not be ignored. For example, if a critical link in an SCADA network has been targeted in the past for an attack, it is likely to be protected via a security control. Let the vulnerability cost matrix of the link between nodes i and j be constructed such that $A_r^{ij} = L_R^{ij}$, and then necessary normalization should be performed when combining the different parameters so that $A_r^{ij} \in [0, 100]$.

6.2.3 History of Communication Links

The introduced link attributes (L_S^{ij}, L_M^{ij}, and L_O^{ij}) may be considered static, while the link history L_H^{ij} is dynamically changing. Here, a dynamical model is proposed for L_H^{ij} to quantify related risk. For simplicity, a Markov model with one step history is adopted. Longer historical couplings are beneficial to tune the Markov model parameters [20].[1]

Let P_A^{ij} denote the probability of "targeting" the communication link (that was not previously targeted) between nodes i and j, and let P_D^{ij} denote the probability that a previously "targeted" communication link that is currently in a "good" state. The "specific actions" that drive these probabilities are not detailed here. Further, let the set of possible Markov model link statuses be $\{\mathcal{G}, \mathcal{T}\}$, and here $\mathcal{G}$ represents a good link while $\mathcal{T}$ indicates a targeted communication link. A finite-state Markov chain process can be characterized using a transition matrix P^{ij} whose

1 Readers should note that this work does not aim to provide or propose intrusion detection capability.

elements provide the likelihood of reaching a future state given the current state. For example, P^{ij} can be represented for a 2-state Markov model as [11]

$$P^{ij} = \begin{bmatrix} 1 - P_A^{ij} & P_A^{ij} \\ P_D^{ij} & 1 - P_D^{ij} \end{bmatrix} \tag{6.4}$$

where

$$\begin{aligned} \mathbb{P}(L_{ij}^{k+1} &= \mathcal{G} \mid L_{ij}^{k} = \mathcal{T}) = P_D^{ij} \\ \mathbb{P}(L_{ij}^{k+1} &= \mathcal{T} \mid L_{ij}^{k} = \mathcal{G}) = P_A^{ij} \\ \mathbb{P}(L_{ij}^{k+1} &= \mathcal{G} \mid L_{ij}^{k} = \mathcal{G}) = 1 - P_A^{ij} \\ \mathbb{P}(L_{ij}^{k+1} &= \mathcal{T} \mid L_{ij}^{k} = \mathcal{T}) = 1 - P_D^{ij} \end{aligned} \tag{6.5}$$

Since the Markov chain process described above is *time homogeneous* [20], the stationary equilibrium and distribution of this Markov chain is considered to derive insights on the long-term likelihoods of certain link states. For the following discussion, the subscript ij is dropped for simplicity. Let the *stationary distribution* of the Markov chain process be described using a row vector as π, which has nonnegative entries that sum to 1. Also, π is unchanged by application of P^{ij}. Thus, π is described using

$$\begin{aligned} 1 &= \sum_{q \in S} \pi_q \\ \pi_q &= \sum_{r \in S} \pi_q P_{qr} \\ 0 &\le \pi_q \le 1 \end{aligned} \tag{6.6}$$

where π_q is the qth element of π where $q \in S$ and $S = \{1, 2\}$ to appropriately span the matrix elements.

Given the described Markov chain process is time homogeneous with finite state space and if $\lim_{n\to\infty} (P^{ij})^n$ exists, the stationary distribution π is calculated from $\lim_{n\to\infty} (P^{ij})^n = \mathbf{1}\pi$ [20]. Thus, if the limit exists, either the transition matrix P^{ij} or the stationary distribution π could be employed to reflect the communication link history L_H^{ij}. Furthermore, given the stationary nature of the probability model, we employ $L_H^{ij} = \pi(1)$, the equilibrium likelihood that the link is in state T (i.e. the communication link between i and j is targeted by an adversary). The above discussion combines the dynamically changing history of communication links with parametric properties to quantify a measure of the link risk L_r^{ij}. Our proposed risk-aware QoS routing will utilize this risk metric as one of the constraints.

A cross-layer risk-aware CPS showing the interplay between the cyber and physical components in a CPS of a smart city infrastructure is illustrated in Figure 6.2.

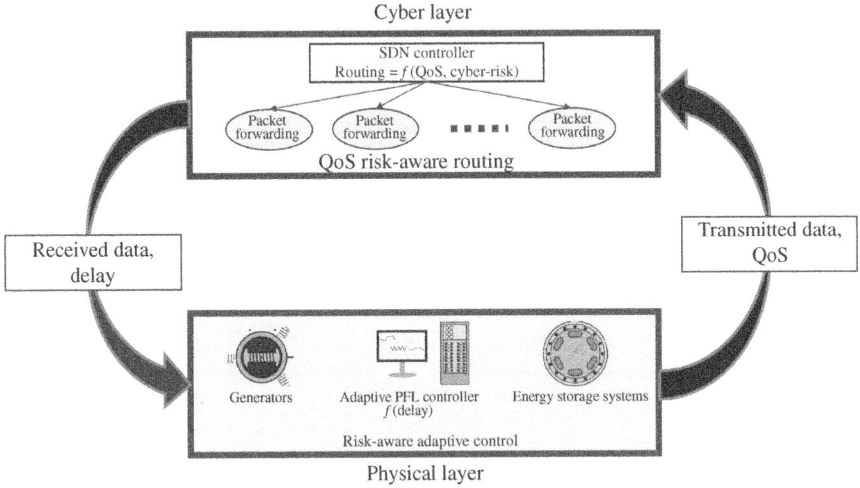

Figure 6.2 Cross-layer risk-aware cyber-physical system.

6.3 Risk-Aware Quality of Service Routing Using SDN

6.3.1 Constrained Shortest Path Routing Problem Formulation

We consider the routing problem between two nodes in a communication network where the route needs to satisfy at least two constraints. Recall the weighted and undirected graph $G(V, E, w)$ that describes the communication network as M-link and N-node networks where $V \in \mathcal{N}^{N \times N}$ and $E \in \mathcal{N}^{M \times M}$. Within this framework, every link $(i, j) \in E$ has two different weights: $d_{ij} > 0$ as a communication delay matrix and $c_{ij} > 0$ as a cost matrix. The routing problem is traditionally concerned with finding the shortest path p with a minimum cost between a source node s and a destination node t.

When employing additional constraints such as delay, a CSP formulation can be used to define the routing problem between the nodes. For source and destination nodes, pair (s, t), $\mathbf{P}_{st}$ denotes the set of paths from s to t. Let also the maximum allowable communication latency in the network be $T_{\max} > 0$. We define the costs associated with any path $p \in \mathbf{P}_{st}$ as

$$\begin{aligned} d(p) &= \sum_{(i,j) \in p} d_{ij} \\ c(p) &= \sum_{(i,j) \in p} c_{ij} \end{aligned} \tag{6.7}$$

We seek through solving the CSP routing problem to find the minimum cost path between source node s and destination node t that satisfies another constraint. Hence, within the proposed QoS constrained routing framework, a feasible path in the solution refers to a routing path that satisfies the maximum allowable network latency $d(p) \leq T_{\max}$. Thus, the CSP problem aims to find the minimum cost feasible path between s and t as $p^* = \arg\min \{c(p)|p \in \mathbf{P}_{st}(T_{\max})\}$ where $\mathbf{P}_{st}(T_{\max})$ is the a set of feasible paths between the two nodes.

An *integer linear program* can be used to further formulate the CSP problem with a set of binary zero/one decision variables x_{ij}. Consequently, for a communication link $(i,j) \in E$, let $x_{ij} = 1$ when the (i,j) link is in path $p_{stand} x_{ij} = 0$ when (i,j) is not. Finding the minimum-cost feasible path is formulated as $\forall(i,j) \in E$ [18, 21, 22]:

$$\begin{aligned} \text{Minimize} \quad & \sum_{(i,j)\in E} c_{ij}\, x_{ij} \\ \text{Subject to} \quad & x_{ij} \in \{0,1\} \\ & \sum_{j\in V} x_{ij} = \sum_{j\in V} x_{ij}, \forall i \in V \backslash \{s,t\} \\ & \sum_{j\in V} x_{sj} = 1 \\ & \sum_{i\in V} x_{it} = 1 \end{aligned} \tag{6.8}$$

The formulation above is an NP-hard problem [21], and different heuristics and approximate algorithms have been proposed for solving (6.8). One approach [23] calculates iteratively the shortest routing path between s and t according to the first metric (i.e. the cost constraint) and then checks if the found path also satisfies the QoS metric (i.e. communication latency).

The Lagrange relaxation-based aggregated cost (LARAC) algorithm [21] was shown to efficiently arrive at the minimum-cost feasible routing path in polynomial time. This relies on using a modified cost function; given $\lambda \geq 0$, the aggregate cost c_λ is defined as

$$c_\lambda(ij) = c_{ij} + \lambda d_{ij} \tag{6.9}$$

For a given λ, a minimal path can easily be calculated. The Lagrange relaxation problem is defined using [21]

$$\mathcal{L}(\lambda) = \min \{c_{\lambda(p)} \; |p \in \mathbf{P}_{st}\} - \lambda T_{\max} \tag{6.10}$$

The above Lagrange relaxation $\mathcal{L}(\lambda)$ ignores the constraints and moves them to the objective function. To arrive at the best lower bound, the dual of the Lagrange relaxation $\mathcal{L}(\lambda)$ is considered as $\mathcal{L}^*$ where

$$\mathcal{L}^* = \max \mathcal{L}(\lambda) \; |\lambda \geq 0 \tag{6.11}$$

Figure 6.3 Pseudocode of the LARAC algorithm.

START LARAC (s, t, d, T)
p_c := Dijkstra Function(s, t, c)
if $(d(p_c) \leq T)$ **then**
  return p_c
end if
p_d := Dijkstra Function(s, t, d)
if $(d(p_d) > T)$ **then**
  return "Solution cannot be found."
end if
loop
  $\lambda := \dfrac{c(p_c) - c(p_d)}{d(p_d) - d(p_c)}$
  r := Dijkstra Function(s, t, c_λ)
  if $(c_\lambda(r) == c_\lambda(p_c))$ **then**
    return p_d
  else if $(d(r) \leq T)$ **then**
    $p_d := r$
  else
    $p_c := r$
  end if
end loop
END ALGORITHM

The work of [21] proved that the λ^* maximizing $\mathcal{L}(\lambda)$ is actually the smallest value at which there exists a routing path with minimal aggregate cost satisfying the delay constraint. Recalling our risk-aware QoS routing problem, the CSP formulation admits the proposed risk metric as the cost metric in addition to the network delay. Thus, the LARAC algorithm shown in Figure 6.3 is adopted as a candidate algorithm. Furthermore, the SDN framework, as will be discussed in the later sections, further enriches the problem by the ability to obtain an updated delay cost matrix of the network.

6.3.2 SDN Architecture and Implementation

Software-defined networking is changing the way that traditional communication networks operate. Traditionally, such networks are composed of large number of devices that communicate and coordinate communication using complex distributed protocols. Network performance management is challenging in traditional networks [24]. The architecture of an SDN can provide manageable and dynamic solutions for adaptable networks. SDN architecture separates traffic routing decisions (the control plane) from simpler traffic forwarding (the data plane). Network applications in an SDN framework are able to obtain detailed traffic statistics from various network devices; thus, an up-to-date global network view can be constructed by such applications. SDN provides an innovative approach for dealing with many challenges in communication and networking of smart grids and smart city infrastructure. Recent works explored potential benefits of SDN in smart grids including [25–31].

An SDN architecture typically includes two main components:

(1) **SDN controller**: A logically centralized function that determines the routing path for network flows.
(2) **SDN switch**: A commodity hardware that is capable of forwarding network flows between source and destination nodes following the SDN controller rules.

Open source tools are adopted for the SDN cyber part of this work. One standard for implementation of an SDN is OpenFlow [32]. Floodlight v1.0 [33] is adopted as the OpenFlow SDN controller. Mininet 2.2.0 [34] can create realistic virtual networks and is adopted for implementing SDN switches. The Floodlight SDN controller communicates via the OpenFlow protocol with the Mininet switches through an abstraction layer that is present at the forwarding hardware. The OpenFlow protocol defines a set of messages that can be exchanged between SDN entities over a secure channel, which facilitates the communication between an SDN controller and its managed SDN switches. Figure 6.5 illustrates the adopted SDN architecture.

6.3.3 Risk-Aware Routing Algorithm

The CSP formulation described above is adopted in this study to model the routing problem that includes a best-effort vulnerable link avoidance while meeting QoS delay constraints. Also, a distributed SDN-based version of the LARAC algorithm is devised here that utilizes the constraints of the risk metric defined in (6.3) and communication delay. The SDN implementation on risk-aware QoS routing is termed as Vulnerable-Link Adaptive-Avoidance (VLAA) routing algorithm. Specifically, a dynamic delay cost matrix is communicated periodically. Communication link delays are observed through the SDN switches at each update interval T_{sd}; if changes in communication delay are observed, the Mininet switches update the OpenFlow Floodlight controller with the new delay cost matrix. In a similar fashion, link risk weights are periodically obtained, and the cost matrix is updated when any changes are observed.

The VLAA algorithm is developed with two components: an SDN switch function is implemented in Mininet using Python, and an SDN control function is implemented in Floodlight using Java. Figure 6.4 shows a flowchart of the implemented VLAA algorithm in the Floodlight controller's side. The following tasks are performed by the algorithm on the controller:

- Listen to messages from the managed SDN switches, calculate link delay value of each communication link, and construct the link-delay cost matrix.

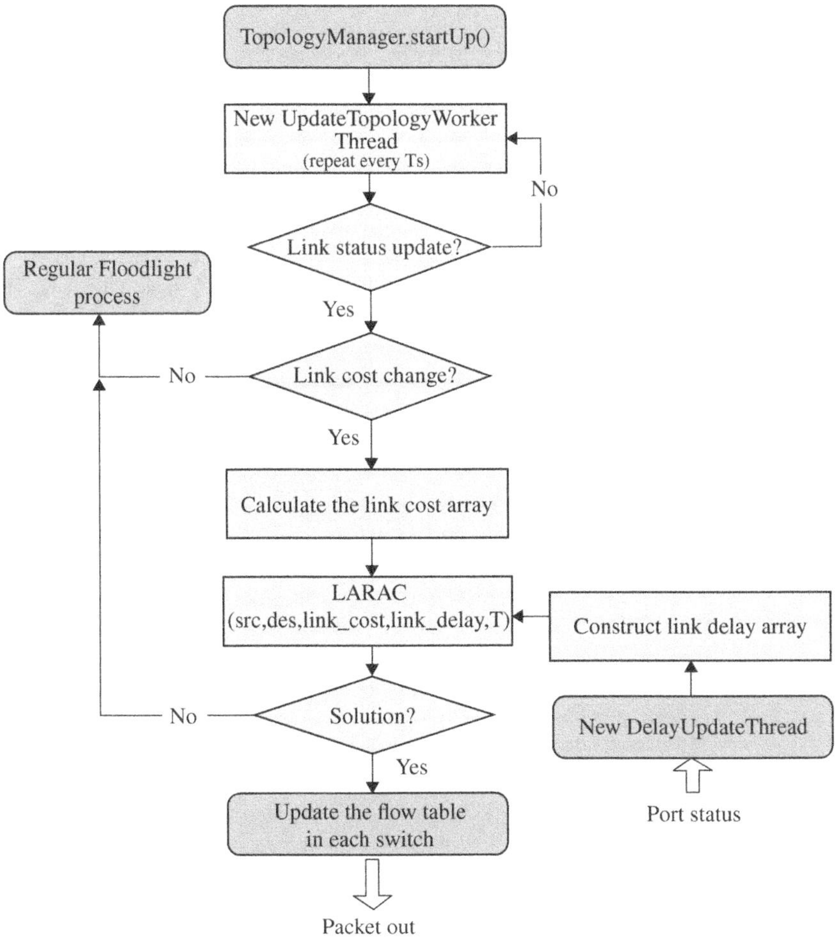

Figure 6.4 Implementation of the SDN controller in the VLAA algorithm.

- Calculate the link-vulnerability cost matrix according to the metrics of (6.3); network operator/managers can modify and calibrate the cost matrix as needed.
- Run a topology-update thread, and check the updated link cost matrix every T_{sd}.
- Calculate the routing paths in the network based on the link cost matrix.
- If the controller detects a change in the cost matrix, it recalculates the routing paths; the SDN controller updates the flow table of the managed switches by advertising a *packet-out* message to each switch.

On the switch's side, the VLAA algorithm collects the values of link latency for the directly connected Mininet switches as demonstrated in Figure 6.6. This function is accomplished using an independent thread for periodically testing the communication link between a switch and all other connected switches

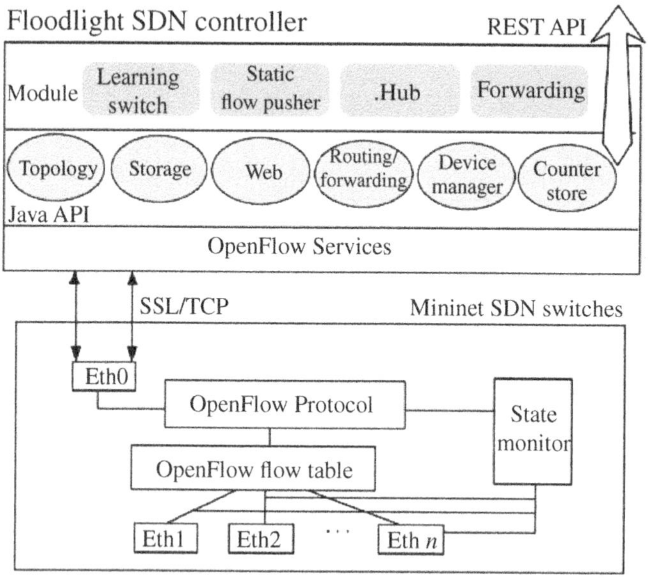

Figure 6.5 SDN architecture.

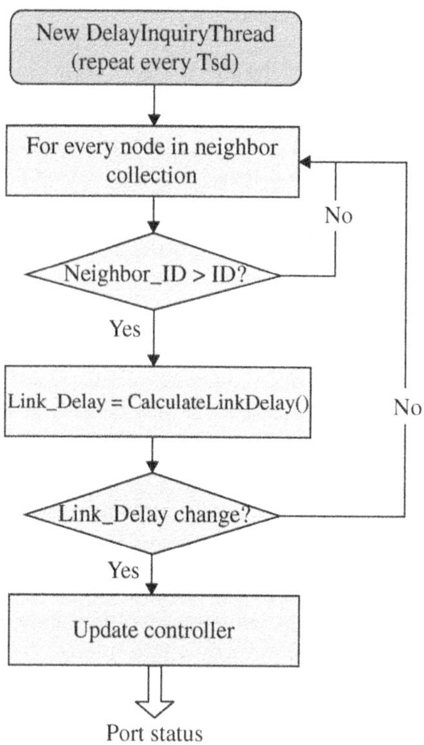

Figure 6.6 Implementation of the SDN switches in the VLAA algorithm.

that have a higher ID. The sampling time can be tuned by the network operator/manager. In the simulation environment in Section 6.5, $T_{sd} = 60$ seconds. Furthermore, link-delay testing is performed three times every T_{sd}, and the average value of those three readings is compared with the last known link-delay value. If the new average delay value is noticeably different from the previous value, the Mininet switch updates the managing SDN controller with the new delay value.

To be compatible with the OpenFlow standard, the Mininet switch exchanges link latency updates with its managing Floodlight controller through *port-status* messages. An SDN switch is expected by OpenFlow standards (v1.0–v1.4) to send *port-status* messages to its managing SDN controller as *port configuration state* changes. Such update events many include any change in *port status* if it was directly brought down by an operator or as specified by the IEEE 802.1D standard.

6.4 Risk-Aware Adaptive Control

6.4.1 Smart Grid Model

The smart grid system is modeled as N-agent CPS. An agent includes a synchronous generator, a sensor, a controller, and an energy storage system (ESS). In this model, a corresponding communication network connects the agents' controllers with sensors. An agent's sensor measures the synchronous generator's rotor speed and angle, and the controller uses the sensor's readings to actuate the ESS through power absorption or injection. In this CPS model, the cyber components include the communication network, controllers, and sensors, and physical components include the traditional power delivery elements and the ESS.

As an example of smart grid system, consider the IEEE 39-bus 10-generator test power system. The swing equation model can be employed to describe the dynamics of the synchronous generators in power systems including the study of transient stability. In this notion, N equals the number of generators in the power system. Furthermore, for generator i where $i \in \{1, \dots, N\}$, E_i is the internal voltage, $P_{e,i}$ is the electrical power, $P_{m,i}$ is the mechanical power, ω_i is the relative and normalized rotor frequency, δ_i is the rotor angle, M_i is the inertia, and D_i is the damping coefficient.

Let $\dot{\omega}_i$ and $\dot{\delta}_i$ denote the time derivatives of ω_i and δ_i, respectively. Therefore, the swing equation model for generator $i, i \in \{1, \dots, N\}$ is defined as [35, 36]

$$\begin{aligned} \dot{\delta}_i &= \omega_i \\ \dot{\omega}_i &= \frac{1}{M_i}(-D_i\, \omega_i + (P_{m,i} - P_{e,i})) \end{aligned} \tag{6.12}$$

and $P_{e,i}$ is expressed using [37]

$$P_{e,i} = \sum_{k=1}^{N} |E_i|\ |E_k|[G_{ik}\cos(\delta_i - \delta_k) + B_{ik}\sin(\delta_i - \delta_k)] \tag{6.13}$$

where, for $i, k \in \{1, \ldots, N\}$, $G_{ik} = G_{ki} \geq 0$ is the Kron-reduced equivalent conductance and $B_{ik} = B_{ki} > 0$ is the Kron-reduced equivalent susceptance between generators i and k. The accelerating power of generator i is $P_{a,i} = P_{m,i} - P_{e,i}$. Thus, the model in (6.12) can be written as

$$\begin{aligned} \dot{\delta}_i &= \omega_i \\ \dot{\omega}_i &= \frac{1}{M_i}[-D_i\ \omega_i + P_{a,i}]\ . \end{aligned} \tag{6.14}$$

6.4.2 Parametric Feedback Linearization Control

Parametric feedback linearization (PFL) is a tunable control scheme that integrates with governor control of synchronous generators. The PFL controller uses system state information to execute a feedback linearization actuation that synchronizes the synchronous generators during system transient instability incidents [38]. This control scheme relies on receiving periodic sensor measurements, and it affects the power system dynamics (if needed) through power absorption or injection via the fast-acting ESS. Through this actuation process, the PFL controller shapes the dynamics of the controlled power system to resemble a series of stable control systems with tunable eigenvalues.

To calculate the control actuation signal, a centralized PFL control scheme receives sensor measurements of the rotor speed and phase angle of all synchronous generators in the power system. The centralized PFL actuation is expressed as [39]

$$u_i = -P_{a,i} - \alpha_i\ \omega_i - \beta_i[\delta_i - \delta_i^*] \tag{6.15}$$

where $\alpha_i \geq 0$ and $\beta_i \geq 0$ are the frequency and phase stability parameters of generator i, respectively, and $\delta^* = [\delta_1^*, \delta_2^*, \ldots, \delta_N^*]^T$ is the desired phase angle of the synchronous generators in the smart grid system. It is observed that $\alpha_i\omega_i$ in u_i will asymptotically drive ω_i of generator i to 0; similarly, $\beta_i(\delta_i - \delta_i^*)$ will settle δ_i of the generators on δ^*. Thus, phase angle cohesiveness and rotor speed synchronization are achieved due to the activation of the PFL control scheme. The centralized PFL control scheme described above in (6.15) cancels the nonlinear terms in the swing equation of generator i in (6.12). As such, after applying the centralized PFL control, the swing equation of the controlled power system reduces into decoupled linear equations.

As observed above, the centralized PFL control relies on obtaining all system measurements as described above. On the other hand, the communication networks and channels that relay the sensor measurements are vulnerable to cyber attacks as described in Section 6.2. Given this limitation, a decentralized PFL control scheme only utilizes the local sensor measurements. The decentralized PFL control signal is expressed as [40]

$$u_i = -\alpha_i\omega_i - \beta_i[\delta_i - \delta_i^*] \tag{6.16}$$

Consequently, $P_{a,i} = P_{m,i} - P_{e,i}$ in (6.13) cannot be estimated or canceled, which results in a partially linearized control system.

6.4.3 Risk-Aware Routing and Latency-Adaptive Control Scheme

Latency tolerance is the longest time delay that a closed-loop control system can tolerate to maintain the stability of the controlled system [41]. A delay-robust PFL control design can improve the controller's latency tolerance as demonstrated in [38]. Define τ as the communication latency between sensors and controllers. A delay-adaptive centralized PFL control scheme is proposed in [38] and can be expressed as

$$u_i(t) = -P_{a,i}(t-\tau) - \alpha_i(\tau)\omega_i(t-\tau) - \beta_i(\tau)[\delta_i(t-\tau) - \delta_i^*] \tag{6.17}$$

A delay-adaptive decentralized PFL control scheme can also be described as

$$u_i(t) = -\alpha_i(\tau)\omega_i(t-\tau) - \beta_i(\tau)[\delta_i(t-\tau) - \delta_i^*] \tag{6.18}$$

Let $\alpha_i^*(\tau), \beta_i^*(\tau)$ denote the values of α_i and β_i of the PFL control in (6.15) and (6.16) that optimize the controller's performance for different values of communication latency. The latency resulting from risk-aware QoS routing between the PFL controller and the sensors is known for each measurement. Consequently, when sensor measurements are received by the PFL controller with a latency of τ, the controller can apply the appropriate values of $\alpha_i^*(\tau), \beta_i^*(\tau)$ given the value of τ. Furthermore, the maximum tolerable control delay would constitute the QoS constraint for the risk-aware routing.

6.5 Simulation Environment and Numerical Analysis

The IEEE New England 39-bus 10-generator test power system is considered. An accompanying communication network topology is proposed to parallel the New England power grid as shown in Figure 6.7. The assumption of installing a network switch at every power system bus is behind the proposed communication network, and consequently a bus is thought to include an SDN forwarding switch. Those switches also conduct traffic measurements that are forwarded to the SDN controller. Furthermore, consider in this communication network a centralized

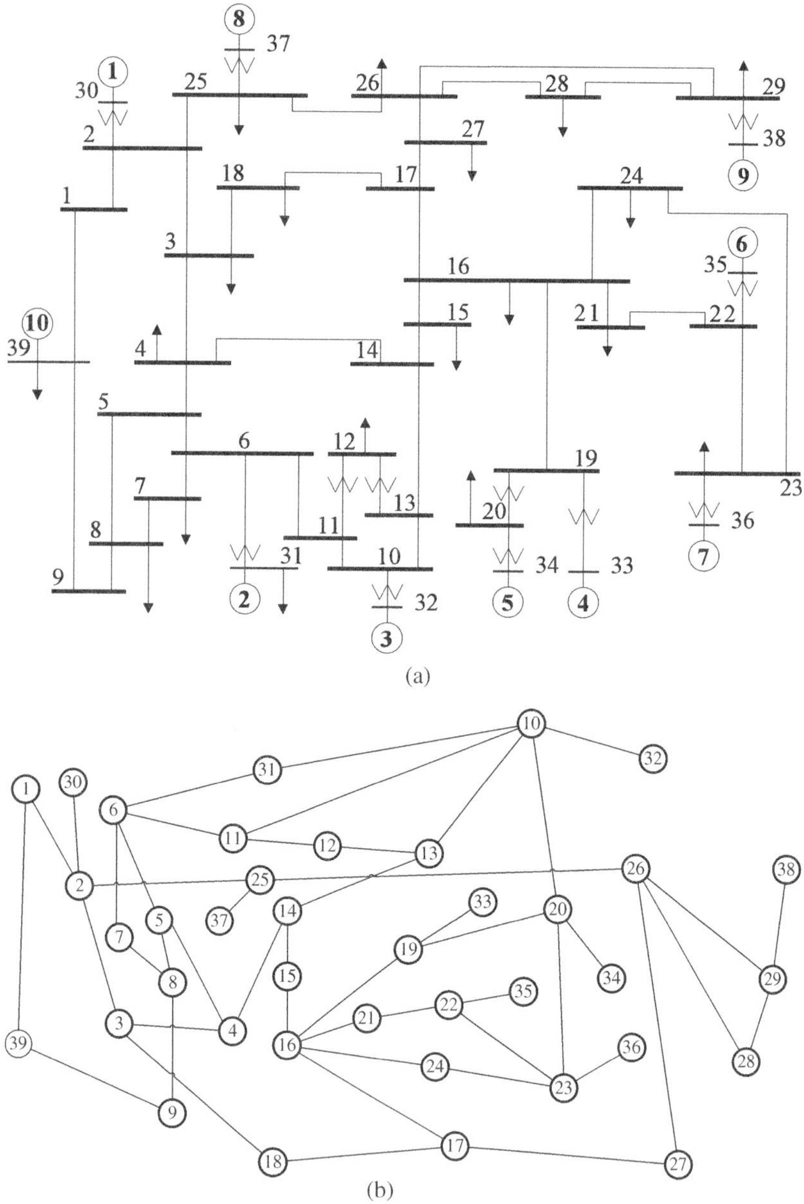

Figure 6.7 Test power system: single line diagram and the accompanying communication network.

SDN controller that computes the forwarding table for an assigned set of SDN switches. To adapt to changing network conditions, the SDN controller applies the received traffic measurements from the switches along with protocol-defined routing rules to dynamically update the forwarding tables of the SDN switches.

The simulation environment runs on a Windows 7 64-bit machine with a 2.53-GHz Intel Core i5 CPU and 8-GB RAM. The simulation environment uses Floodlight v1.0 [33] to implement SDN controllers and Mininet 2.2.0 [34] to implement the SDN switches. Moreover, iPerf [42] and JPerf [43] are used for collecting network performance statistics. JPerf is the graphical user interface (GUI) front end of iPerf that is a network tool possibly used to create Transmission Control Protocol (TCP) and User Datagram Protocol (UDP) data streams and measure the throughput in a communication network that carries TCP and UDP traffic.

Risk-aware QoS routing is considered in the numerical analysis since it is expected that the VLAA routing algorithm might not be able to completely avoid high-risk communication links to satisfy the QoS constraints. As such, some of the network traffic is expected to be routed through high-risk links if they are unavoidable. To compare the effectiveness of the different routing algorithms, we therefore compute a *traffic leakage* metric by how much traffic passes through a certain vulnerable link as a measure of the algorithm's avoidance to that link.

The VLAA algorithm implementation uses the built-in Dijkstra algorithm in Floodlight, where the cost parameters are specified as integers. Thus, communication link cost and latency matrices are defined as integer matrices; this is accomplished by scaling a parameter's non-integer value by a multiplier. In addition, the constraints and metrics of the LARAC algorithm need to be normalized to achieve a fair comparison between different constraints. This normalization is required to bring the two metrics to same levels through a preprocessing stage. To do that, the range of value in the metrics is set to be in the [0–100] range. Consequently, the desired normalization can be achieved through linear interpolation.

6.5.1 Avoiding Vulnerable Communication Links While Meeting QoS Constraint

In this simulation, we study a case where some vulnerable communication links are distributed along the network. The performance of three algorithms is compared: VLAA algorithm, static Dijkstra algorithm, and the dynamic delay Dijkstra (denoted as DD Dijkstra) algorithm when the link latency matrix is used as a dynamics link cost parameter. As mentioned above, some of these vulnerable links could have undesirable historical records, may belong to other service providers, or are having less secure wireless links. Thus, A_v will show higher values for the more vulnerable communication links.

The communication network is randomly assigned risk metrics for each communication link. For randomly -selected (s, t) pairs, different communication paths are enumerated. The accompanying grid includes 38 nodes; thus, a total of 1444 paths (38 nodes × 38 nodes) can be established. Out of the 1444 paths, 278 paths are found to be different between Dijkstra and VLAA algorithms. Also, between DD Dijkstra and VLAA algorithms, there are 254 paths found to be different.

Furthermore, to compare the performance of the VLAA algorithm with that of Dijkstra and DD Dijkstra algorithms, traffic leakage metric is calculated. Figure 6.8 shows the packet leakage in vulnerable communication links for the three routing algorithms. It is shown that the proposed VLAA routing algorithm achieves lower packet leakage while meeting QoS (i.e. communication latency) constraints. It is to be noted though that the VLAA algorithm's performance is slightly affected in terms of bandwidth and latency because the algorithm could choose longer communication paths to avoid high-vulnerable links. Figure 6.9 illustrates algorithm performance in terms of end-to-end network latency and bandwidth using box plots[2]. It is observed from the results that the VLAA algorithm achieves lower leakage while meeting the QoS requirements. Similarly, the VLAA algorithm's delay performance is slightly affected because to avoid high-vulnerability communication links, the algorithm could choose longer paths.

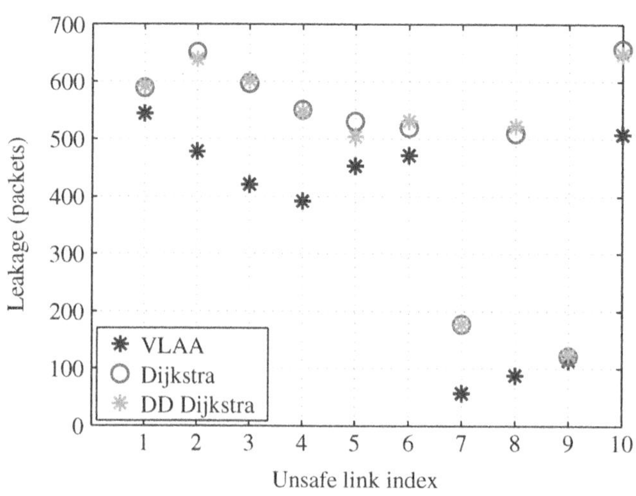

Figure 6.8 Packets transmitted on high-risk links.

2 Box plots provide standardized approach of displaying data distribution based on the five-number summary: minimum, first quartile, median, third quartile, and maximum [44].

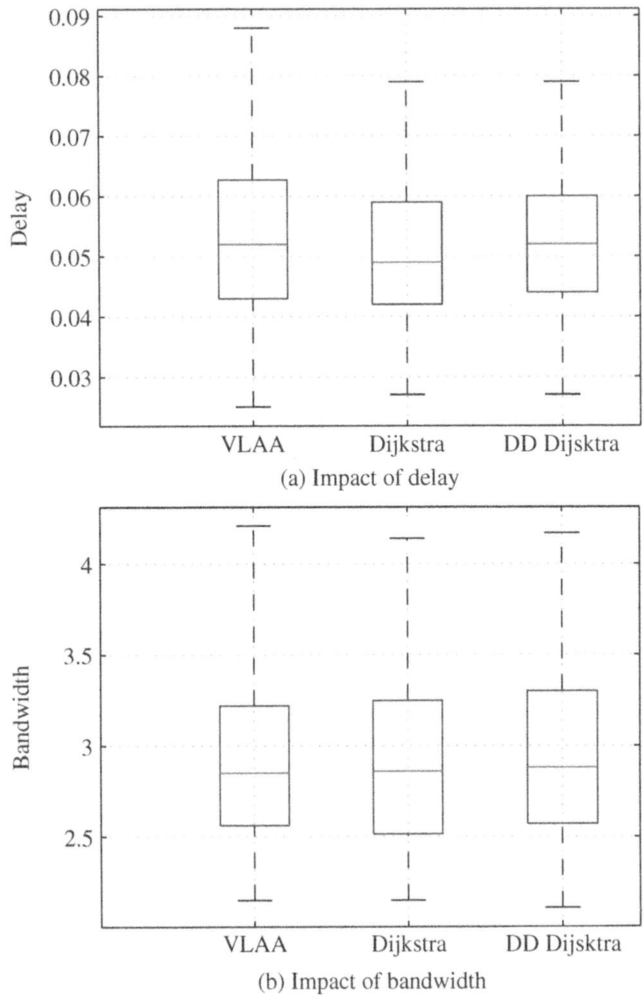

Figure 6.9 Impact of communication delay and bandwidth on performance.

6.5.2 Algorithm Overhead Comparison

The average processing time of communication route calculations is tested for the following test power systems [45]: IEEE 118-bus power system, IEEE 57-bus power system, IEEE 39-bus power system, and WSCC 9-bus power system. The goal is to compare the performance and time overhead between the three algorithms.

In this experiment, cost of different communication links is randomly sampled from a [1–150] range, and the processing time is averaged over 10 random (s, t)

Table 6.1 Processing time comparison.

	Processing time (second)		
Test system	**VLAA algorithm**	**Dijkstra algorithm**	**Dynamic delay Dijkstra**
IEEE 118-bus power system	26.622	26.617	26.620
IEEE 57-bus power system	22.854	22.848	22.852
IEEE 39-bus power system	20.149	20.145	20.147
IEEE 14-bus power system	15.518	15.512	15.513
WSCC 9-bus power system	13.124	13.118	13.121

pairs. In Table 6.1, it is shown that the processing time overhead of the VLAA algorithm is very low compared with the Dijkstra algorithms.

6.5.3 Impact of QoS Constraints

The impact of the smart grid's traffic having different classes of QoS on the performance of the VLAA algorithm is investigated. The impact of varying the value of T_{max} on traffic leakage and on algorithm overhead metrics is studied and demonstrated in Figure 6.10. Increased leakage is observed for lower values of T_{max}; this is explained by noting that when QoS constraints become tighter (which can be translated into requiring lower values of T_{max}), the set of feasible routing paths becomes smaller, and more leakage occurs. In addition, the results in Figure 6.10b show that more conservative QoS metrics (i.e. lower values of T_{max}) lead to more algorithm processing time. This result can be explained by noting that the number of iterative calls inside the LARAC portion of the VLAA algorithm in Figure 6.3 increases to arrive at feasible paths for (s, t). However, no noticeable difference in processing time is observed between the VLAA and Dijkstra algorithms when the T_{max} requirement is relaxed.

6.5.4 Impact on Distributed Control

We investigate the impact of the VLAA algorithm on physical PFL distributed control. The IEEE 39-bus test power system (along with the corresponding communication network) is initially in a normal operating state, and then a DoS attack is assumed to target the cyber network for a duration of two seconds. The average latency of the communication system is shown in Figure 6.11 for both Dijkstra and VLAA algorithms. Furthermore, a physical attack (three-phase fault at Bus 5)

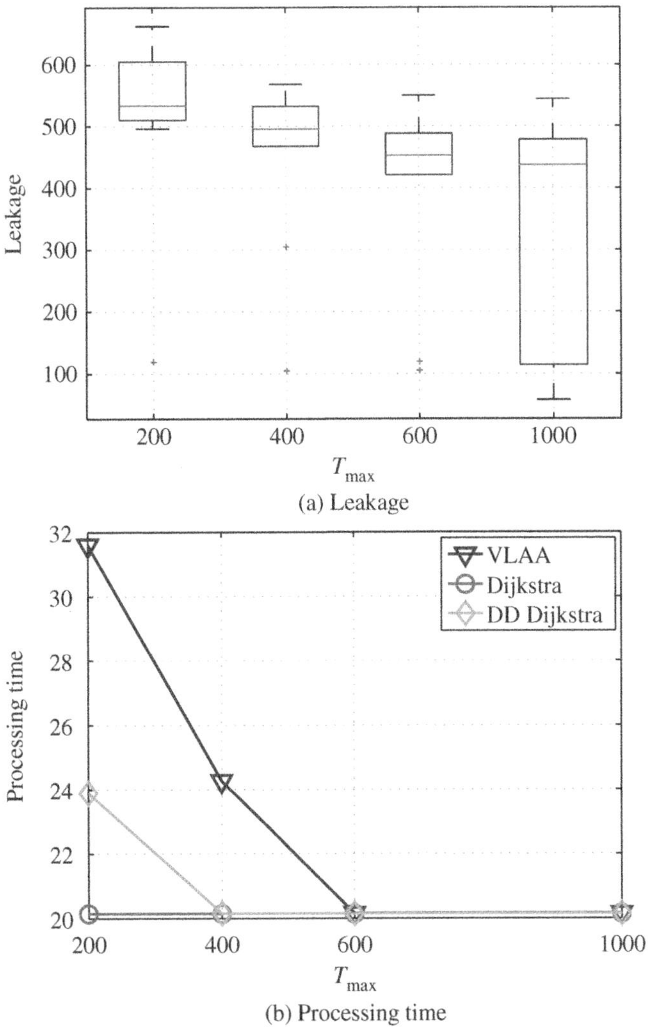

Figure 6.10 Leakage and processing time versus maximum allowable network delay (T_{max}).

occurs for 100 ms during the cyber attack. The delay-adaptive PFL distributed control scheme [39, 40] is activated to counter the physical fault and bring the power system to stability. Compared with its performance under the Dijkstra algorithm, the PFL controller requires 3% less time to stabilize the power system under the VLAA algorithm. Further studies are needed to investigate this approach in larger-scale power systems where benefits could be improved further.

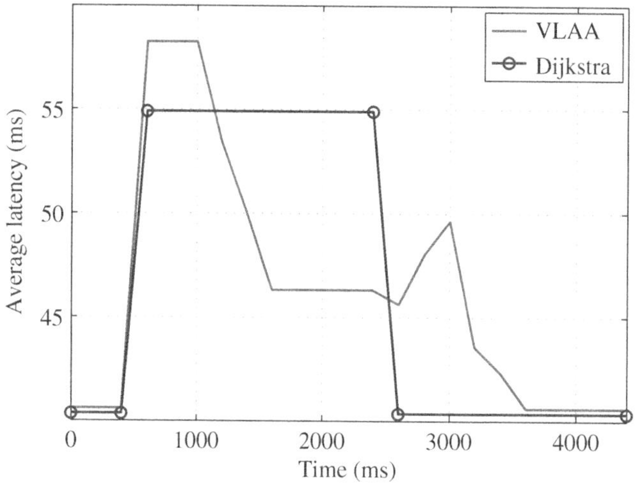

Figure 6.11 Cyber attack impact on average latency.

6.6 Conclusions

CPS in smart cities have been targeted by advanced persistent threats, and current approaches to detection, response, and recovery of cyber incidents need to evolve. In this chapter, we propose a cross-layer risk-aware control framework to enhance the cyber-physical resilience of the smart city infrastructure against cyber and physical disturbances and risks. We consider the communication networks in smart city infrastructure and specifically focus on developing a framework for enhancing communication security in a smart city while maintaining QoS requirements.

Our proposed approach considers utilizing an adaptive risk-aware routing on the cyber layer that takes into account performance constraints of the physical system as additional constrains. We utilize SDN technology framework for the mitigation of cyber risks while meeting the essential control QoS requirements in CPS within smart city infrastructure. The SDN controller is able, through this formulation, to maintain awareness of the various cyber risks in the CPS, thus optimizing communication routing within the network based on specific metrics of interest. For practical reasons, the communication delay between the data source and destination is considered as the QoS metric. The SDN routing problem is treated using double CSP formulation with additional constraints that quantify cyber risks in communications links.

On the physical layer, our approach utilizes a risk-aware adaptive controller. Specifically, a parametric feedback linearization control is adopted that modulates risk as uncertainty and utilizes additional energy storage resources to improve system resilience during and after physical disturbances. Using its situational awareness of risks through the smart city systems, the proposed cyber-physical control framework integrates both cyber and physical layers of control to mitigate risks and respond to disturbances.

Benefits of the proposed approach in enhancing the cyber-physical resilience of the smart city infrastructure are showcased through examples focusing on smart grid applications. Demonstrated through different smart grid test cases, the proposed risk-aware cyber-physical control framework is able to mitigate communication link risks, meet QoS requirements, and achieve transient stability.

References

1 Carter, C.N.A. and Rieti, J. (2021). Sidewalk Labs cancels plan to build high-tech neighbourhood in Toronto amid COVID-19. (accessed January).

2 Cranley, B.I.E. (2021). 8 Cities that have been crippled by cyberattacks–and what they did to fight them. (accessed January).

3 Li, H. and Zhang, W. (2010). QoS routing in smart grid. *IEEE Global Telecommunications Conference (GLOBECOM)*, pp. 1–6.

4 Zaballos, A., Vernet, D., and Selga, J.M. (2013). A genetic QoS-aware routing protocol for the smart electricity networks. *International Journal of Distributed Sensor Networks* 9 (9). 1–12.

5 Pavlidou, F.-N. and Koltsidas, G. (2008). Game theory for routing modeling in communication networks-a survey. *Journal of Communications and Networks* 10 (3): 268–286.

6 Shah, G., Gungor, V.C., Akan, O.B. et al. (2013). A cross-layer QoS-aware communication framework in cognitive radio sensor networks for smart grid applications. *IEEE Transactions on Industrial Informatics* 9 (3): 1477–1485.

7 Zhu, Q., Wei, D., and Basar, T. (2011). Secure routing in smart grids. *Workshop on Foundations of Dependable and Secure Cyber-Physical Systems (FDSCPS)*, pp. 55–59.

8 Srikantha, P. and Kundur, D. (2015). Denial of service attacks and mitigation for stability in cyber-enabled power grid. *Innovative Smart Grid Technologies Conference (ISGT), 2015 IEEE Power & Energy Society*, pp. 1–5.

9 Wei, J. and Kundur, D. (2016). GOAlie: Goal-seeking obstacle and collision evasion for resilient multicast routing in smart grid. *IEEE Transactions on Smart Grid*. 7 (2): 567–579.

10 Zhao, J., Hammad, E., Farraj, A., and Kundur, D. (2016). Network-aware QoS routing for smart grids using software defined networks. In: *Smart City 360*, 384–394. Springer.

11 Hammad, E., Zhao, J., Farraj, A., and Kundur, D. (2016). Mitigating link insecurities in smart grids via QoS multi-constraint routing. *2016 IEEE International Conference on Communications Workshops (ICC)*, IEEE, pp. 380–386.

12 Farraj, A., Hammad, E., and Kundur, D. (2017). On the use of energy storage systems and linear feedback optimal control for transient stability. *IEEE Transactions on Industrial Informatics* 13 (4): 1575–1585.

13 Farraj, A., Hammad, E., and Kundur, D. (2017). On the impact of cyber attacks on data integrity in storage-based transient stability control. *IEEE Transactions on Industrial Informatics.* 13 (6): 3322–3333.

14 Egilmez, H.E., Dane, S.T., Bagci, K.T. et al. (2012). OpenQoS: An openflow controller design for multimedia delivery with end-to-end quality of service over software-defined networks. *Signal & Information Processing Association Annual Summit and Conference (APSIPA ASC), 2012 Asia-Pacific*, IEEE, pp. 1–8.

15 Ishimori, A., Farias, F., Cerqueira, E., and Abelém, A. (2013). Control of multiple packet schedulers for improving QoS on OpenFlow/SDN networking. *2013 2nd European Workshop on Software Defined Networks (EWSDN)*, pp. 81–86.

16 Network Working Group (1990). RFC 1195 use of OSI IS-IS for routing in TCP/IP and dual environments. www.ietf.org/rfc/rfc1195.txt (accessed December 2020).

17 Network Working Group (1998). RFC 2328 OSPF version 2. www.ietf.org/rfc/rfc2328.txt (accessed December 2020).

18 Xiao, Y., Thulasiraman, K., Xue, G., and Jüttner, A. (2005). The constrained shortest path problem: algorithmic approaches and an algebraic study with generalization. *AKCE International Journal of Graphs and Combinatorics* 2 (2): 63–86.

19 I. I. S. Forum (2021). Information Risk Assessment Methodology 2 (IRAM2). (accessed January 2021).

20 Norris, J.R. (1998). *Markov Chains, Cambridge Series in Statistical and Probabilistic Mathematics.* Cambridge University Press.

21 Jüttner, A., Szviatovski, B., Mécs, I., and Rajkó, Z. (2001). Lagrange relaxation based method for the QoS routing problem. *20th Annual Joint Conference of the IEEE Computer and Communications Societies (INFOCOM)*, Volume 2, 859–868.

22 Kuipers, F., Van Mieghem, P., Korkmaz, T., and Krunz, M. (2002). An overview of constraint-based path selection algorithms for QoS routing. *IEEE Communications Magazine* 40 (12): 50–55.

23 Lee, W. and Hluchyj, M. (1994). Multi-criteria routing subject to resource and performance constraints. *ATM Forum* 94: 0280.

24 Nunes, B., Mendonca, M., Nguyen, X.-N. et al. (2014). A survey of software-defined networking: past, present, and future of programmable networks. *IEEE Communication Surveys and Tutorials* 16 (3): 1617–1634.

25 Cahn, A., Hoyos, J., Hulse, M., and Keller, E. (2013). Software-defined energy communication networks: from substation automation to future smart grids. *IEEE International Conference on Smart Grid Communications (SmartGrid-Comm)*, pp. 558–563.

26 Kim, Y.-J., He, K., Thottan, M., and Deshpande, J.G. (2014). Virtualized and self-configurable utility communications enabled by software-defined networks. *IEEE International Conference on Smart Grid Communications (SmartGrid-Comm)*, pp. 416–421.

27 Zhang, J., Seet, B.-C., Lie, T.-T., and Foh, C.H. (2013). Opportunities for software-defined networking in smart grid. *International Conference on Information, Communications and Signal Processing (ICICS)*, pp. 1–5.

28 Molina, E., Jacob, E., Matias, J. et al. (2015). Using software defined networking to manage and control IEC 61850-based systems. *Computers and Electrical Engineering* 43: 142–154.

29 Dong, X., Lin, H., Tan, R. et al. (2015). Software-defined networking for smart grid resilience: opportunities and challenges. *ACM Workshop on Cyber-Physical System Security*, pp. 61–68.

30 Sydney, A., Ochs, D.S., Scoglio, C. et al. (2014). Using GENI for experimental evaluation of software defined networking in smart grids. *Computer Networks* 63: 5–16.

31 Goodney, A., Kumar, S., Ravi, A., and Cho, Y.H. (2013). Efficient PMU networking with software defined networks. *IEEE International Conference on Smart Grid Communications (SmartGridComm)*, pp. 378–383.

32 Open Networking Foundation. (2012). Software-defined networking: the new norm for networks. *ONF White Paper*.

33 Project Floodlight (2015). Project Floodlight. www.projectfloodlight.org/floodlight/ (accessed December 2020).

34 Mininet (2015). Mininet. www.mininet.org/ (accessed December 2020).

35 Anderson, P. and Fouad, A. (1994). *Power System Control and Stability, IEEE Power Systems Engineering Series*. IEEE Press.

36 Dörfler, F. and Bullo, F. (2010). Synchronization and transient stability in power networks and non-uniform Kuramoto oscillators. *American Control Conference (ACC)*, June/July 2010, pp. 930–937.

37 Bergen, A. and Vittal, V. (2000). *Power Systems Analysis*, 2e. Prentice-Hall.

38 Farraj, A., Hammad, E., and Kundur, D. (2015). A systematic approach to delay-adaptive control design for smart grids. *2015 IEEE International Conference on Smart Grid Communications (SmartGridComm)*, pp. 768–773.

39 Farraj, A., Hammad, E., and Kundur, D. (2015). A cyber-enabled stabilizing controller for resilient smart grid systems. *ISGTInnovative Smart Grid Technologies (ISGT)*, February 2015, pp. 1–5.
40 Hammad, E., Farraj, A., and Kundur, D. (2015). A resilient feedback linearization control scheme for smart grids under cyber-physical disturbances. *ISGTInnovative Smart Grid Technologies (ISGT)*, February 2015, pp. 1–5.
41 Wu, H., Tsakalis, K.S., and Heydt, G.T. (2004). Evaluation of time delay effects to wide-area power system stabilizer design. *IEEE Transactions on Power Systems* 19: 1935–1941.
42 iPerf.fr. Iperf. www.iperf.fr/ (accessed December 2020).
43 xjperf (2015). xjperf. code.google.com/p/xjperf/ (accessed December 2020).
44 csbsju.edu. Box Plot: Display of Distribution. www.physics.csbsju.edu/stats/box2.html (accessed December 2020).
45 Information Trust Institute (2015). Power Flow Cases Illinois Center for a smarter Electric Grid. publish.illinois.edu/smartergrid/power-cases/ (accessed December 2020).

7

Wind Speed Prediction Using a Robust Possibilistic C-Regression Model Method: A Case Study of Tunisia

Achraf J. Telmoudi, Moez Soltani, Lotfi Chaouech, and Abdelkader Chaari

The National Higher Engineering School of Tunis (ENSIT), Department of Electrical Engineering, Laboratoire d'Ingenierie des Systemes Industriels et des Energies Renouvelables, University of Tunis, Montfleury, Tunis BP 56-1008, Tunisia

7.1 Introduction

Tunisia has been trying to reduce its dependence on fossil fuel markets (in particular, oil and gas) by diversifying its sources of energy in the last few years. Strategically, the energy independence is the key factor of sovereignty and overall national security in the long term. According to International Renewable Energy Agency (IRENA) [1], the demand for electricity in Tunisia is expected to increase by 6% per year and thus exceed 20 TWh by 2020 and 33 TWh by 2030. Based on the report made by the Tunisian Intended Nationally Determined Contribution (INDC) [2], the energy consumption of the building sector in particular will increase by a factor of almost 2.5 between 2010 and 2030 and will represent the highest share of energy consumption from 22% in 2010 to 35% in 2030. Moreover, according to the National Institute of Statistics, Tunisia's energy deficit alone accounted for 32.4% of the overall deficit of the country's trade balance in 2018, which weakens the state's capacity to support economic growth and therefore job creation and social services. To these challenges, we must add the climatic constraints, which will force us to deviate from the traditional ways of energy consumption and development and to maintain sustained economic growth and to reduce CO_2 emissions. Tunisia proposes reducing its greenhouse gas emissions across all sectors by 41% in 2030 according to INDC's forecasts [2].

This situation implies for Tunisia major strategic and socioeconomic challenges. Between deficit and compensation, the energy bill is quite substantial in the state budget. To satisfy the energy requirement, the development of new energy sources has been pointed out, in particular the wind energy, which seems to be one of

Cyberphysical Smart Cities Infrastructures: Optimal Operation and Intelligent Decision Making, First Edition. Edited by M. Hadi Amini and Miadreza Shafie-khah.

the most promising ones with a very high international growth rate. In Tunisia, the potential for wind energy production remains untapped with current power capacity of 245 MW [3, 4], despite that many studies [5–7] have been shown that Tunisia has a huge available potential of wind and can be considered suitable for the development of the offshore projects. Moreover, wind measurements, in the North Tunisia, revealed wind speeds of 7–15 m/s, indicating strong potential for wind power development [7, 8].

Wind energy has recently experienced unprecedented growth worldwide. The growing importance of integrating wind energy into the electrical system is driving researchers to develop more reliable techniques for forecasting wind energy. However, wind energy production is extremely intermittent as it is influenced by a number of extremely volatile factors such as weather conditions and wind speed. Moreover, before connecting any source of decentralized energy production to the electricity grid, it is necessary to check the conditions of connection (same frequency, same voltage, same phase), which is very difficult with the wind systems given the random production of energy and forecasting. Since wind energy penetrates significantly into the electricity grid, it becomes essential to predict accurately the wind energy production. For this, in this chapter, we propose a new methodology for wind speed prediction of three Tunisian meteorological sites: Bizerte, Nabeul, and Monastir. Indeed, fuzzy c-regression model (FCRM) [9] algorithms have been shown to be more effective in many real nonlinear system identification applications [10–15]. However, all the solutions proposed on these FCRM clustering-based methods only focus on the improvement of the clustering quality without taking into consideration the data distribution and/or the outliers (measurement uncertainty and environmental noise). As a result, it is often difficult to obtain a high-quality approximation of real systems. To improve the quality of FCRM, a robust possibilistic c-regression model (RPCRM) clustering technique is applied for finding and constructing a model from wind speed data sets, which would be the most efficient and would yield the smallest errors by comparing the predicted and actual models. Finally, we compared the results of the predicted wind speed for different seasons (winter, spring, summer, and autumn) of four sites to check if the results obtained are correct.

This chapter is organized as follows. Section 7.2 discusses in more detail the theoretical concept of the proposed method and presents the wind speed data collection. In Section 7.3, performance improvements of the proposed method will be given to verify the effectiveness for modeling and prediction of the wind speed. The main contributions of this study are discussed in Section 7.4.

7.2 Data Collection and Method

This section describes the data set used and the method applied for a robust prediction of wind direction in details.

7.2.1 Data Description

The data sets of wind speed are collected from three local meteorological sites in the Tunisia: Bizerte, Nabeul, and Monastir. The geographic localization of three local meteorological sites is shown in Figure 7.1. Our sample consists of 365 historical wind speed data and obtained from the observations of the 12 months, from 1 January to 30 December 2018 at each site. Table 7.1 shows the elevation above sea level, location, and population for the three local sites.

7.2.2 Robust Possibilistic C-Regression Models

The RPCRM was proposed by Soltani et al. [16] and Telmoudi et al. [17] to take into account the presence of noise or outliers in the data set. This is a generalized algorithm for special cases of FCRM [9] and possibilistic c-regression model [18] algorithms. It minimizes the following objective function:

$$J(U,\theta) = \sum_{k=1}^{N}\sum_{i=1}^{c}(\mu_{ik})^m(E_{ik}(\theta_i))^2 + \sum_{k=1}^{N}\delta^2(\mu_{*k})^m \tag{7.1}$$

with

$$E_{ik}(\theta_i) = |y_k - [\mathbf{x}_k \quad 1].\ \theta_i^T| \tag{7.2}$$

Here $\mu_{*k} = 1 - \sum_{i=1}^{c}\mu_{ik}$ is the membership of point x_k in the noise cluster, and the noise distance δ can be calculated as follows:

$$\delta^2 = \gamma_i, \text{ for } i \text{ such that } \max_{1\le i\le c}\{\mu_{ij}\} \tag{7.3}$$

with

$$\gamma_i = \frac{\sum_{k=1}^{N}\mu_{ik}^m\ E_{ik}^2}{\sum_{k=1}^{N}\mu_{ik}^m} \tag{7.4}$$

Using the Lagrange multiplier method and differentiating Eq. (7.1) with respect to μ_{ij}, the membership function update equation is

$$\mu_{ik} = \frac{\left(\frac{1}{E_{ik}}\right)^{\frac{2}{m-1}}}{\sum_{j=1}^{c}\left(\frac{1}{E_{jk}}\right)^{\frac{2}{m-1}} + \left(\frac{1}{\delta}\right)^{\frac{2}{m-1}}} \tag{7.5}$$

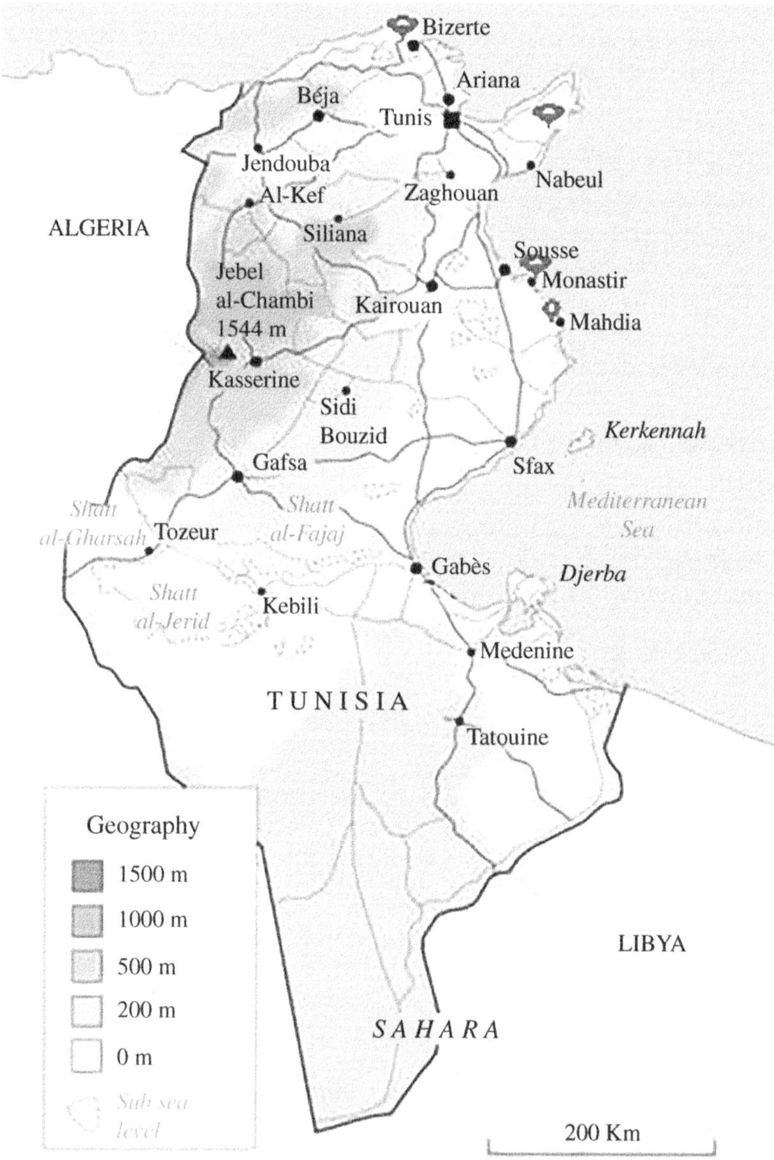

Figure 7.1 Geographic localization of three local meteorological sites.

Table 7.1 Performance indexes.

Sites	Elevation (m)	Location	Population
Bizerte	54	37.28°N 9.87°E	115 268
Nabeul	4	36.46°N 10.74°E	56 387
Monastir	11	35.67°N 10.75°E	71 546

From Eq. (7.5), we can obtain also the following update equation:

$$\mu_{ik} = \left\{ \frac{\left(\frac{1}{E_{ik}}\right)^{\frac{2}{m-1}}}{\sum_{j=1}^{c}\left(\frac{1}{E_{jk}}\right)^{\frac{2}{m-1}}} \right\} \left\{ \frac{\sum_{j=1}^{c}\left(\frac{1}{E_{jk}}\right)^{\frac{2}{m-1}}}{\sum_{j=1}^{c}\left(\frac{1}{E_{jk}}\right)^{\frac{2}{m-1}} + \left(\frac{1}{\delta}\right)^{\frac{2}{m-1}}} \right\} \tag{7.6}$$

and then it can be rewritten as

$$\mu_{ik} = \left\{ \frac{\left(\frac{1}{E_{ik}}\right)^{\frac{2}{m-1}}}{\left(\frac{1}{d_k}\right)^{\frac{2}{m-1}}} \right\} \left\{ \frac{\left(\frac{1}{d_k}\right)^{\frac{2}{m-1}}}{\left(\frac{1}{d_k}\right)^{\frac{2}{m-1}} + \left(\frac{1}{\delta}\right)^{\frac{2}{m-1}}} \right\} \tag{7.7}$$

with $d_k^2 = \left(\sum_{i=1}^{c} \frac{1}{E_{ik}^2}\right)^{-1}$ is a modified version of harmonic mean distance.

Finally, the membership function of RPCRM is defined as follows:

$$\begin{aligned} \mu_{ik} &= \left\{ \frac{\left(\frac{1}{E_{ik}}\right)^{\frac{2}{m-1}}}{\left(\frac{1}{d_k}\right)^{\frac{2}{m-1}}} \right\} . \left(1 + \left(\frac{d_k}{\delta}\right)^{\frac{2}{m-1}}\right)^{-1} \\ &= \mu_{ik}^{\text{FCRM}} . \mu_{ik}^{\text{PCRM}} \end{aligned} \tag{7.8}$$

The first term on the left-hand side of Eq. (7.8) is similar to the original FCRM objective function, whereas the second term is due to the possibilistic C-regression model (PCRM) clustering extension.

The RPCRM clustering algorithm is summarized in six steps:

Step 1. Fix the number of clusters c, set $m > 1$, and select the initial parameter vectors θ_i. Fix the stopping criterion $\epsilon > 0$, and initialize $U^{(0)}$ (e.g. random).

Repeat for $r = 1, 2, \ldots$.

Step 2. Compute $\theta_i^{(r)}$ using Eq. (7.9).

Step 3. Compute the error measure $E_{ik}(\theta_i^{(r)})$ via Eq. (7.2).

Step 4. Calculate δ^2 via Eq. (7.3).

Step 5. Update $U^{(r)}$ using Eq. (7.8).

Step 6. Compute $err = ||U^{(r)} - U^{(r-1)}||$.

Until $err \leq \epsilon$, then stop. Otherwise set $r = r + 1$ and return to Step 2.

The weighted least squares (WLS) method can be applied to estimate the consequent parameters for each rule, which is described as follows:

$$Y = \begin{bmatrix} y_1 \\ y_2 \\ \vdots \\ y_M \end{bmatrix}, \quad U_i = \begin{bmatrix} u_i(\mathbf{x}_1) & 0 & \ldots & 0 \\ 0 & u_i(\mathbf{x}_2) & \ldots & 0 \\ \vdots & \vdots & \ddots & \vdots \\ 0 & 0 & \ldots & u_i(\mathbf{x}_M) \end{bmatrix}$$

$$\theta_i = \left[\overline{X}^T U_i \overline{X}\right]^{-1} \overline{X}^T U_i Y \tag{7.9}$$

where $\overline{X} = [\mathbf{x}_k \quad 1]$, $\mathbf{x}_{kj}$ is jth coordinate of the vector $[\mathbf{x}_k \quad 1]$ $(j = 1, 2, \ldots, M + 1)$ and $U = \{\mu_{ik}, i = 1, \ldots, c\}$ is the partition matrix of the membership degrees.

The fuzzy sets centers $\mathbf{v}_{ik}$, and the respective standard deviations $\boldsymbol{\sigma}_{ik}$ of premise part are defined as [19]

$$\mathbf{v}_{ij} = \frac{\sum_{k=1}^{N} \mu_{ik} \mathbf{x}_{kj}}{\sum_{k=1}^{N} \mu_{ik}}, \quad i = 1, 2, \ldots, c; \quad j = 1, 2, \ldots, M \tag{7.10}$$

$$\boldsymbol{\sigma}_{ij} = \sqrt{\frac{2\sum_{k=1}^{N} \mu_{ik} \ (\mathbf{x}_{kj} - \mathbf{v}_{ij})^2}{\sum_{k=1}^{N} \mu_{ik}}} \tag{7.11}$$

7.2.3 Wind Speed Data Analysis Procedure

The design procedure for wind speed prediction is as illustrated in Figure 7.2.

7.3 Experiment and Discussion

In this section, the methodology developed in Section 7.1 is applied not only to validate the performance and effectiveness of RPCRFM algorithm but also to model and estimate the wind speed. In this chapter, four performance indices are defined: mean square error (MSE), normalized root mean square error (NRMSE), mean absolute percentage error (MAPE), and coefficient of determination R^2:

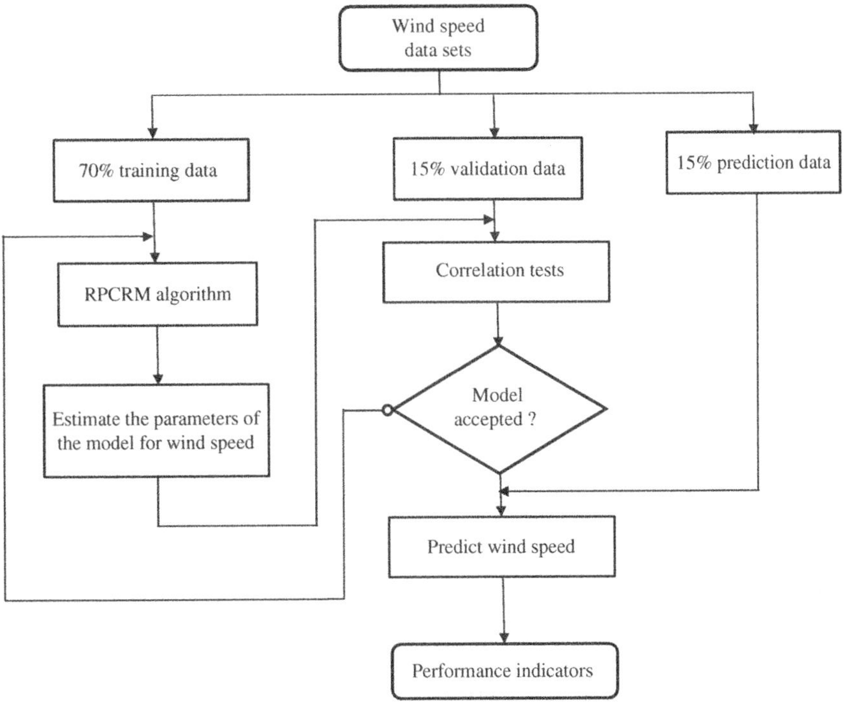

Figure 7.2 General procedure for wind speed prediction.

$$\text{MSE} = \frac{1}{N}\sum_{k=1}^{N}\left(y_k - \hat{y}_k\right)^2 \tag{7.12}$$

$$\text{MAPE} = \frac{1}{N}\sum_{k=1}^{N}\left|\frac{y_k - \hat{y}_k}{y_k}\right| \times 100 \tag{7.13}$$

$$\text{NRMSE} = \frac{\sqrt{\frac{1}{N}\sum_{k=1}^{N}\left(y_k - \hat{y}_k\right)^2}}{y_{k,\max} - y_{k,\min}} \tag{7.14}$$

$$R^2 = 1 - \frac{\sum_{k=1}^{N}\left(y_k - \hat{y}_k\right)^2}{\sum_{k=1}^{N}\left(y_k - \bar{y}\right)^2} \tag{7.15}$$

where y is the actual output, $\hat{y}$ is the model output, $\bar{y}$ is the mean of y, and N is the number of data. The wind speed data is first decomposed into three subsets. The first subset is used for training, the second subset is used to evaluate the final model, and the last one is used to predict wind speed. It is approximated by the following fuzzy model structure:

$$\hat{y}(k) = f\left(y(k-1), y(k-2), y(k-3), y(k-4)\right) \tag{7.16}$$

where k denotes the sampling instant and f is an unknown relationship approximated by the RPCRM algorithm.

The comparison between predicted model and real model for training data set and validation data set of the different local sites is presented in Figures 7.3–7.5 using the RPCRM method with the following settings: $c = 10$, $m = 2$, $\epsilon = 0.0001$.

Clearly it can be seen that the proposed procedure can successfully identify the experimental data sets as shown by Figures 7.3–7.5. Moreover, the speed wind errors between the real and the estimated model based on the performance indexes listed in Table 7.2 is relatively small, which shows that the algorithm is effective.

To further examine the performance of the proposed model, the model validity test is applied to check whether the model fits the data adequately without any biased. Based on the findings reported in [20] and [21], the identified model is adequate if the following correlation tests are satisfied:

$$\begin{aligned} \Phi_{\zeta'\zeta'}(\tau) &= E\left[\left(\xi(k) - \overline{\xi(k)}\right)\left(\xi(k-\tau) - \overline{\xi(k)}\right)\right] = \delta(\tau) \\ \Phi_{\zeta'(\zeta^{2'})}(\tau) &= E\left[\left(\xi(k) - \overline{\xi(k)}\right)\left(\xi^2(k-\tau) - \overline{\xi^2(k)}\right)\right] = 0 \\ \Phi_{\zeta^{2'}\zeta^{2'}}(\tau) &= E\left[\left(\xi^2(k) - \overline{\xi^2(k)}\right)\left(\xi^2(k-\tau) - \overline{\xi^2(k)}\right)\right] = \delta(\tau) \end{aligned} \tag{7.17}$$

with $\zeta = y - \hat{y}$, ϕ is cross-correlation function, the overbar is the mean value, and δ is the Kronecker delta.

The correlation functions Eq. (7.17) should remain inside the 95% confidence intervals. The 95% confidence intervals for large N are approximately

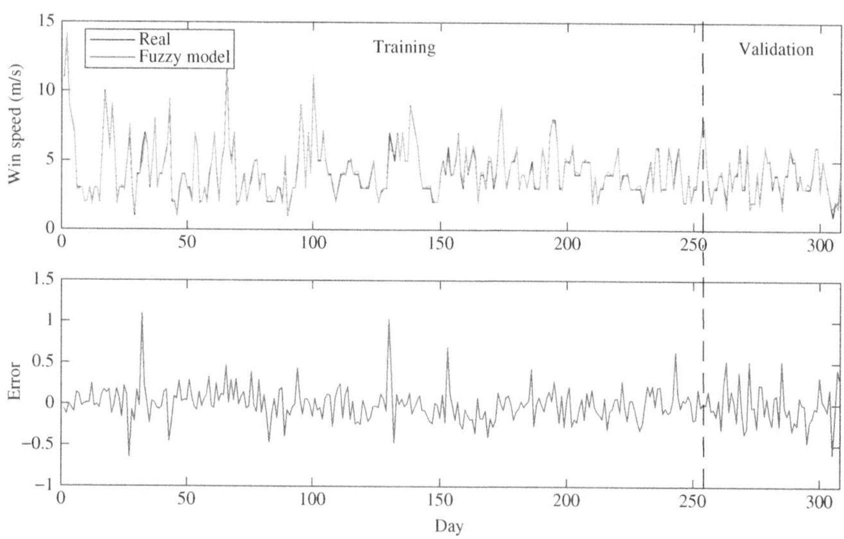

Figure 7.3 Training and validation results for Bizerte wind speed.

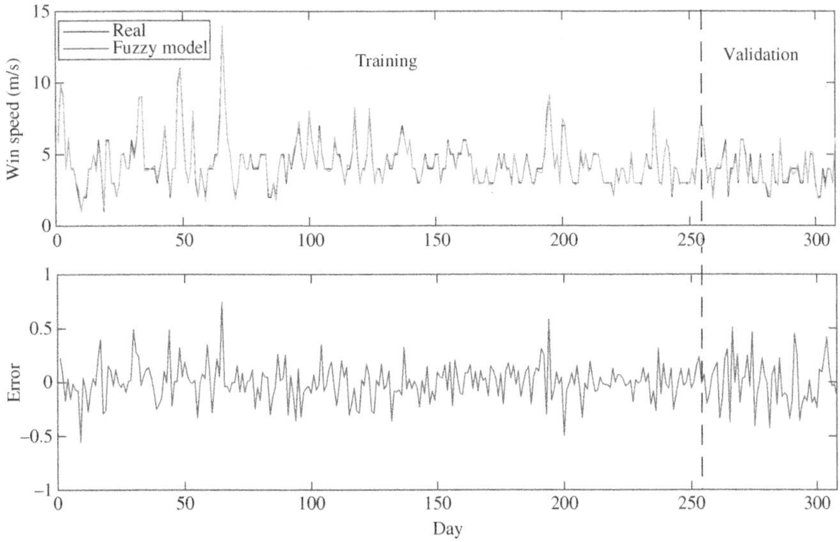

Figure 7.4 Training and validation results for Nabeul wind speed.

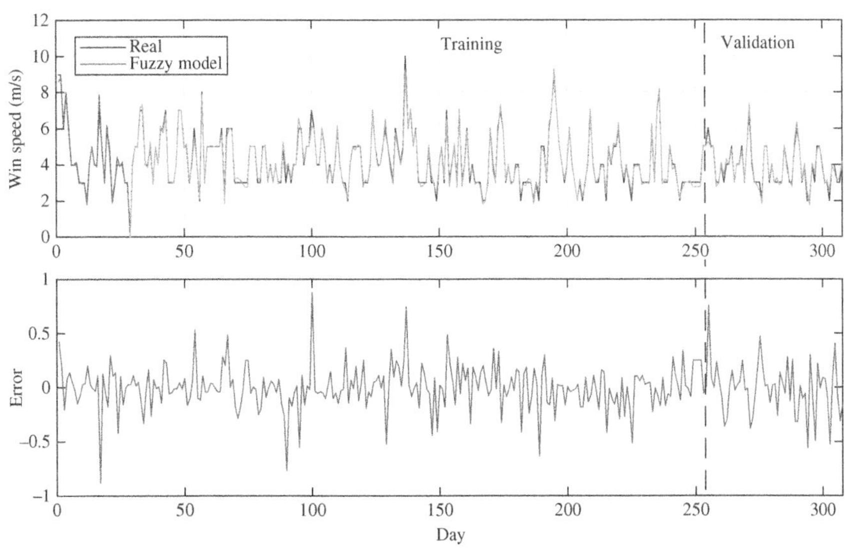

Figure 7.5 Training and validation results for Monastir wind speed.

Table 7.2 Performance indexes of the three local sites.

		MSE	MAPE (%)	NRMSE	R^2
Bizerte	Identification	0.0397	3.6943	0.0153	0.9983
	Validation	0.0585	7.0293	0.0484	0.9961
Nabeul	Identification	0.0281	3.0540	0.0129	0.9988
	Validation	0.0512	5.3258	0.0453	0.9969
Monastir	Identification	0.0417	3.7671	0.0204	0.9979
	Validation	0.0606	5.4474	0.0492	0.9961

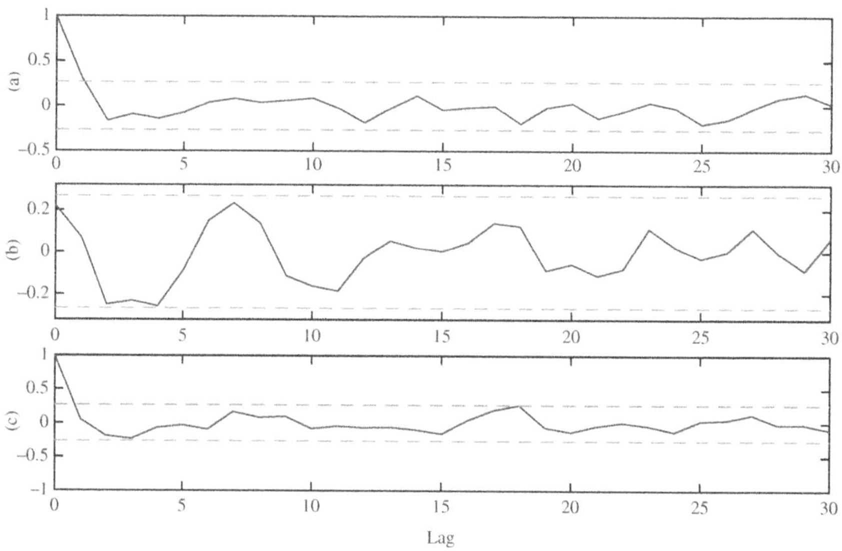

Figure 7.6 Model validity tests of the model for Bizerte wind speed: (a)$\Phi_{\zeta'\zeta'}$, (b) $\Phi_{\zeta'(\zeta^{2'})}$, and (c) $\Phi_{\zeta^{2'}\zeta^{2'}}$.

$\pm 1.96/\text{sqrt}(N)$ [21]. The model validity tests are plotted in Figures 7.6–7.8. These plots confirm that the RPCRM model is statistically valid and suitable to use for predicting the wind directions. Combining the performance index values in Table 7.3 with the prediction results plotted in Figures 7.9–7.11, we can be seen that the prediction results obtained for three sites exhibit a good agreement between the model's outputs and the real outputs.

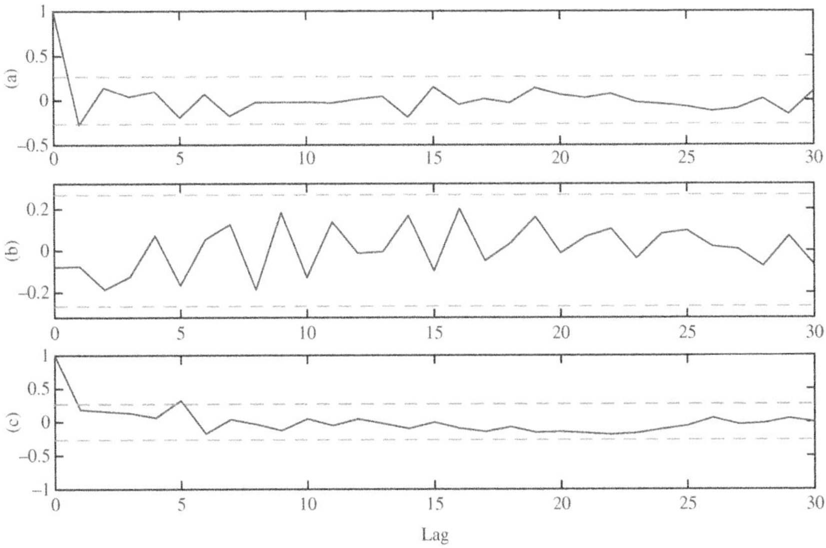

Figure 7.7 Model validity tests of the model for Nabeul wind speed : (a)$\Phi_{\zeta'\zeta'}$, (b) $\Phi_{\zeta'(\zeta^{2'})}$, and (c) $\Phi_{\zeta^{2'}\zeta^{2'}}$.

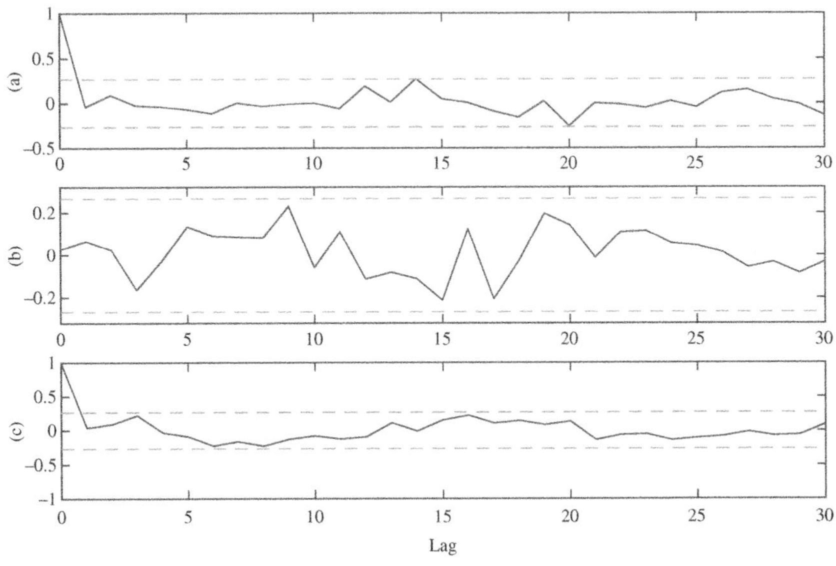

Figure 7.8 Model validity tests of the model for Monastir wind speed: (a)$\Phi_{\zeta'\zeta'}$, (b) $\Phi_{\zeta'(\zeta^{2'})}$, and (c) $\Phi_{\zeta^{2'}\zeta^{2'}}$.

Table 7.3 Prediction performance indexes of the three local sites.

	MSE	MAPE (%)	NRMSE	R^2
Bizerte	0.1904	10.2492	0.0485	0.9923
Nabeul	0.1083	6.3203	0.0329	0.9959
Monastir	0.0908	6.2606	0.0335	0.9960

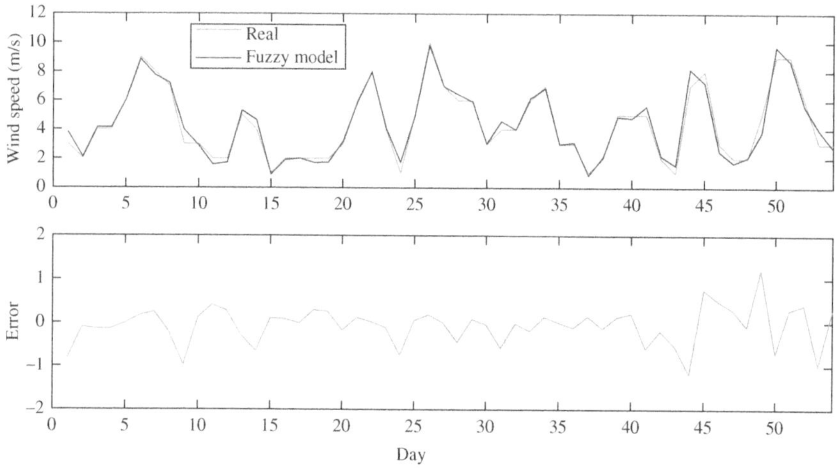

Figure 7.9 Prediction results for Bizerte wind speed.

To give more confidence to the RPCRM models obtained of the three local sites, we will use the last subset that we have not used either during the training learning phase or during the validation phase. The performance of our fuzzy models to predict the wind speed of the different sites is shown in Figures 7.9–7.11.

7.4 Conclusion

In this chapter, a new methodology based on an RPCRM has been designed and presented for simultaneously model parameter estimation and wind speed prediction in Tunisia. The application of the proposed methodology to real data shows that the method gives good results and can achieve a high estimation accuracy in terms of the MSE, root mean square percentage error, MAPE, and coefficient of determination. However, the proposed method also requires a lot of work in the model development. As future work, we intend to enlarge the data set by taking into account other factors such as wind speed, temperature, and pressure.

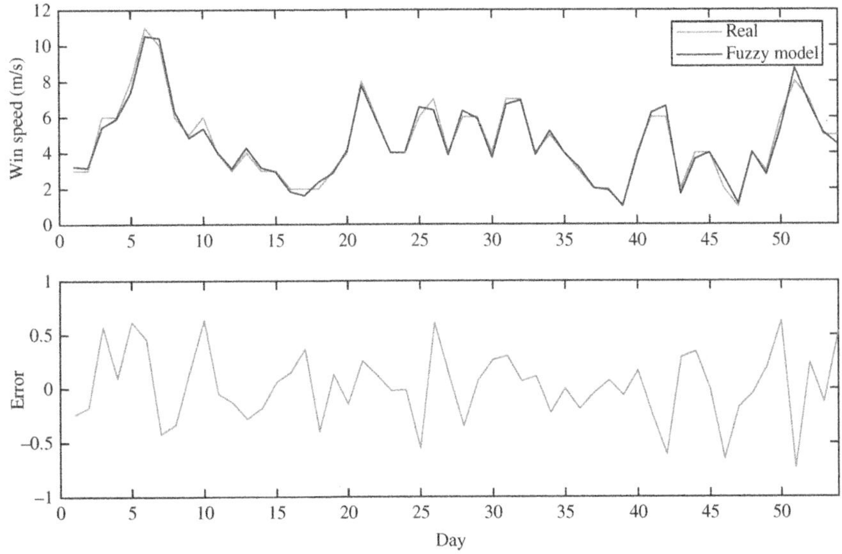

Figure 7.10 Prediction results for Nabeul wind speed.

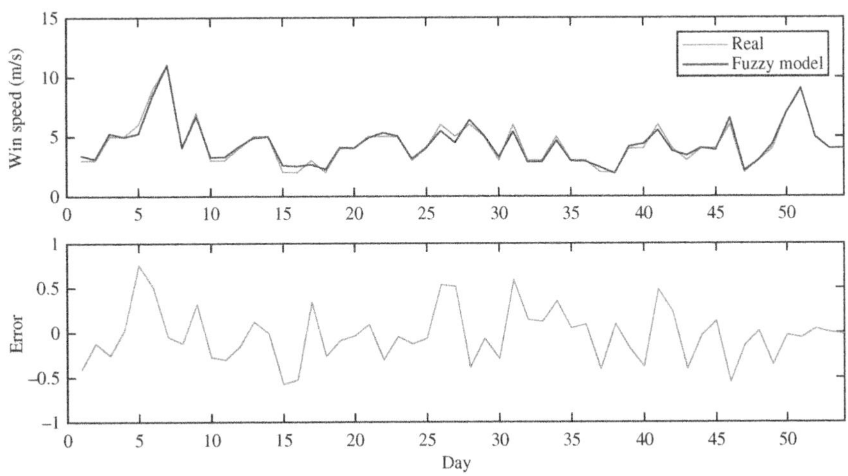

Figure 7.11 Prediction results for Monastir wind speed.

References

1 EIB (2015). Evaluating Renewable Energy Manufacturing Potential in the Mediterranean Partner Countries. *Tech. rep.* European Investment Bank. https://www.eib.org/attachments/femip_study_evaluating_renewable_energy_manufacturing_potential_en.pdf.
2 ANME (2015). NAMA for renewable energy and energy efficiency in the building sector in Tunisia. Mitigation Momentum.
3 Laure, D. (2016). Country Profile on Tunisia. *Tech. rep.* Renewable Energy Solutions for the Mediterranean. https://www.res4med.org/wp-content/uploads/2017/11/Country-Profile-Tunisia-Report_05.12.2016.pdf.
4 Salman, Z. (2020). Solar Energy Prospects in Tunisia. *Tech. rep.* EcoMENA. https://www.ecomena.org/solar-tunisia/.
5 Elamouri, M. and Amar, F.B. (2008). Wind energy potential in Tunisia. *Renewable Energy* 33 (4): 758–768.
6 Maatallah, T., El Alimi, S., Dahmouni, A.W., and Nasrallah, S.B. (2013). Wind power assessment and evaluation of electricity generation in the gulf of Tunis, Tunisia. *Sustainable Cities and Society* 6: 1–10.
7 Florin, O., Deleanu, L., Rusu, L., and Georgescu, C. (2016). Evaluation of the wind energy potential along the Mediterranean Sea coasts. *Energy Exploration & Exploitation* 34 (5): 766–792.
8 Dahmouni, A., Salah, M.B., Askri, F. et al. (2011). Assessment of wind energy potential and optimal electricity generation in Borj-Cedria, Tunisia. *Renewable and Sustainable Energy Reviews* 15 (1): 815–820.
9 Hathaway, R.J. and Bezdek, J.C. (1993). Switching regression models and fuzzy clustering. *IEEE Transactions on Fuzzy Systems* 1 (3): 195–204.
10 Li, C., Zhou, J., Chang, L. et al. (2017). T–S fuzzy model identification based on a novel hyperplane-shaped membership function. *IEEE Transactions on Fuzzy Systems* 25 (5): 1364–1370. http://dx.doi.org/10.1109/TFUZZ.2016.2598850.
11 Zou, W., Li, C., and Zhang, N. (2018). AT–S fuzzy model identification approach based on a modified inter type-2 FRCM algorithm. *IEEE Transactions on Fuzzy Systems* 26 (3): 1104–1113. http://dx.doi.org/10.1109/TFUZZ.2017.2704542.
12 Taieb, A., Soltani, M., and Chaari, A. (2018). A fuzzy c-regression model algorithm using a new PSO algorithm. *International Journal of Adaptive Control and Signal Processing.* 32: 115–133.
13 Telmoudi, A.J., Soltani, M., and Chaari, A. (2018). Identification of PH neutralization process based on a modified adaptive fuzzy c-regression algorithm. http://dx.doi.org/10.1109/icosc.2018.8587628.

14 Li, C., Zou, W., Zhang, N., and Lai, X. (2018). An evolving T–S fuzzy model identification approach based on a special membership function and its application on pump-turbine governing system. *Engineering Applications of Artificial Intelligence* 69: 93–103.

15 Telmoudi, A.J., Soltani, M., Belgacem, Y.B., and Chaari, A. (2020). Modeling and state of health estimation of nickel–metal hydride battery using an EPSO-based fuzzy c-regression model. *Soft Computing* 24 (10): 7265–7279.

16 Soltani, M., Telmoudi, A.J., Chaouech, L., and Chaari, A. (2017). Robust possibilistic c-regression models algorithm. 2017 *4th International Conference on Control, Decision and Information Technologies (CoDIT)*, IEEE, pp. 0466–0470.

17 A. J. Telmoudi, M. Soltani, L. Chaouech, A. Chaari (2020b)Parameter estimation of nonlinear systems using a robust possibilistic c-regression model algorithm, *Journal of Systems Science and Systems Engineering*. 234 (1). 134–143.

18 Kung, C.-C. and Ku, H.-C. (2014). TS fuzzy affine linear modeling algorithm by possibilistic c-regression models clustering algorithm. *2014 IEEE International Conference on Fuzzy Systems (FUZZ-IEEE)*, IEEE, pp. 1242–1247.

19 Soltani, M., Chaari, A., and Hmida, F.B. (2012). A novel fuzzy c-regression model algorithm using a new error measure and particle swarm optimization. *International Journal of Applied Mathematics and Computer Science* 22 (3): 617–628.

20 Mendez, E. and Billings, S.A. (2001). An alternative solution to the model structure selection problem. *IEEE Transactions on Systems, Man, and Cybernetics Part A: Systems and Humans* 31 (6): 597–608.

21 Billings, S.A. (2013). Model validation. In: *Nonlinear System Identification*, 119–147. Wiley. http://dx.doi.org/10.1002/9781118535561.ch5.

8

Intelligent Traffic: Formulating an Applied Research Methodology for Computer Vision and Vehicle Detection

Gabrielle Bakker-Reynolds, Emre Erturk, and Istvan Lengyel

Eastern Institute of Technology, School of Computing, Hawke's Bay, New Zealand

8.1 Introduction

8.1.1 Introduction

As the potential benefits of smart city services continue to attract attention, focus is placed on what new instruments and approaches may be deployed to achieve better results. Examining innovative technologies draws attention to computer vision, a field of computer science that enables computers to process, perceive, and identify images, mimicking how humans make sense of the physical world through different learning architectures [1].

In the recent years, one area of interest for computer vision is the ability to support traffic management systems. Since computer vision holds the capabilities to classify and detect objects, people, and visual details in real time, the ways in which it may assist the development of traffic management solutions are vast. Current examples of computer vision-driven software solutions can be demonstrated within travel assistance and navigation, parking management and enforcement, real-time traffic control, automated vehicle identification, and license plate recognition [2]. In traffic management solutions, computer vision tools can be used in conjunction with the Internet of things (IoT). However, the core city management objectives should generally determine their strategic purpose.

In the context of this research, computer vision is important for object detection. Object detection can be understood as a technique that enables software solutions to detect semantic objects of certain classes within images, video files, or real-time recording [3]. Established domains of object detection are demonstrative in software solutions harnessing facial detection and pedestrian detection capabilities. Subsequently, this research sets out to investigate how computer vision can be

Cyberphysical Smart Cities Infrastructures: Optimal Operation and Intelligent Decision Making, First Edition. Edited by M. Hadi Amini and Miadreza Shafie-khah.

utilized to support a traffic management prototype by providing a cost-effective and efficient way to support traffic flow analysis.

8.1.2 Background

This research aims to develop a vehicle detection prototype for traffic flow analysis by harnessing the capabilities of computer vision. The function of computer vision within this context is to introduce an innovative method that can be utilized to gain insight into busy and peak traffic times. These insights can then be used for a variety of motivations, such as enabling planning and development of new routes and assisting in scheduling maintenance on particular streets or roads.

Presently, the primary way traffic flow analysis is conducted in the selected region of New Zealand through the employment of pneumatic tube-based sensors alongside MetroCount traffic counters [4]. The pneumatic tube-based sensors are placed alongside the roads, and the counter module registers the pulses produced by each vehicle as they drive over the tubes [4]. Additionally, reliance is placed on people to monitor an intersection for a specific period of time. However, this approach leaves room for human error, can take up a lot of time, and, overall, may be considered a mundane task to carry out. Hence, it is significant that the proposed vehicle detection prototype is optimized to accurately detect the number of vehicles. Due to the significance of accurate vehicle detection, an evaluation tool is employed to carry out assessments. Consequently, evaluative measures are set in place to ensure that the accuracy and function of the prototype are analyzed accordingly.

8.1.3 Problem Statement

8.1.3.1 Purpose of Research

This research is concerned with examining how computer vision can be used for vehicle detection. The purpose of this is to serve as an innovative alternative to current traffic flow analysis techniques.

Another goal of this research is to know how innovative technologies may be used to contribute to increase the efficiency of current traffic management approaches in smart cities. This is because in the recent years, computer vision has been noted as an effective and innovative tool in aiding many different areas of general traffic management [5, 6]. Furthermore, a prototype can be designed for capturing traffic flow data, to be evaluated with regard to accuracy and time.

8.1.3.2 Research Questions

The primary research question is as follows:

1. How can computer vision be used to support traffic flow analysis?

Accompanying questions of this research include the following:

2. What is required within the planning, development, and implementation phases of a prototype that performs accurate vehicle detection?
3. How will the performance of a vehicle-detection prototype be measured?
4. What are the barriers to the development of a vehicle-detection prototype?

8.1.3.3 Study Aim and Objectives

The central aim of the research is to investigate and discuss a vehicle-detection prototype. Based on this aim, the three central objectives are as follows:

1. To describe the planning, development, and implementation phases of a modern vehicle-detection prototype.
2. To propose a modern computer vision-driven prototype that can perform accurate vehicle detection for traffic flow analysis.
3. To understand how to assess the performance of a prototype based on evaluation criteria.

8.1.3.4 Significance and Structure of the Research

This research offers a unique perspective and serves as a contribution to the small amount of existing literature covering the role of computer vision-driven traffic solutions within New Zealand and other countries. The significance of this research is the way it informs of how such a traffic monitoring solution might be developed, integrated, and customized. The research also draws attention to a current traffic flow analysis implementation to highlight what changes may be made to strengthen existing processes.

Furthermore, this research explores how incorporating emerging technologies, specifically computer vision-based approaches within traffic flow analysis in New Zealand, can offer affordances. The significance of this comes from noting how computer vision is explored and employed on a global scale. For instance, successful use cases abroad include the use of computer vision to automate traffic lights to decrease congestion and save costs [7], assess the retroreflectivity of road signs to save time and manual maintenance [8], and estimate traffic flow through automation, minimizing the margin of error and offering higher rates of accuracy [9]. Thus, successful use cases such as the ones highlighted above inform on how computer vision is one of the most popular emerging technologies used within traffic management. However, looking into past usage of computer vision illustrates very few use cases, suggesting that there may be barriers preventing further usage.

This research reports on the investigation, development, and testing and trialing of a computer vision-driven prototype to perform vehicle detection – with some limitations. The first limitation to note is the lack of information reporting on the specifications of the current traffic flow analysis approaches employed. Although it is apparent what tools are implemented to carry out traffic flow analysis within New Zealand, not having the definite information reporting on factors such as accuracy, time efficiency, and cost makes it difficult to compare the research's prototype against it. The second potential limitation has to do with the current pandemic circumstances. COVID-19 may impact the research by limiting access to resources required to carry out the research projects. One example of this could be through long wait periods when purchasing required products online from overseas vendors.

Section 8.2 presents a literature review covering relevant academic research within the field of computer vision. Section 8.3 is composed of the methodology portion of the research, addressing the uses of a design-based research approach and examining the proposed study design and the software and hardware used. This also introduces the methodology including the design and development, testing, analysis, and redesign of the prototype. References are made to the literature, and various insights, benefits, and potential barriers are discussed. Additionally, an outline surrounding the next stages of the research project is presented to be continued in Chapter 9. The last section is the conclusion, which also offers directions for future research.

8.2 Literature Review

8.2.1 Introduction

The literature review seeks to create familiarity with the use of computer vision and its present role within traffic management systems. The sources examined throughout the literature review only include primary, peer-reviewed sources available from the databases such as Google Scholar, ProQuest, and ERIC and are divided into thematic subheadings. Keywords used to select the sources included "*computer vision*," "*computer vision and traffic management*," "*machine learning*," "*deep learning*," "*benefits of computer vision*," "*object detection in computer vision*," "*object recognition in computer vision*," and "*technical issues in computer vision*."

8.2.2 Machine Learning, Deep Learning, and Computer Vision

8.2.2.1 Machine Learning

Machine learning (ML) can be understood as a field of computer science that focuses on the study of computer algorithms to carry out intelligent predictions based on a given data input [10]. As ML algorithms are given more data, they are able to perform better and change or alter how they process data over time by providing feedback on previous performances. This, in turn, serves as an extension off of traditional statistical modeling approaches [11]. Additional reasons for the growing interest in ML include advancements in big data, significant increase in computational power, and large progress in the general development and design of ML algorithms [12].

Some of the most popular and widely used ML algorithms fall under either the categories of supervised learning or unsupervised learning. Supervised learning techniques focus on either classification to predict a data label by mapping inputs to outputs or regression, with the objective to predict a data quantity [13]. Conversely, unsupervised learning techniques focus on providing either clustering, which concentrates on finding any fundamental groupings among data, or dimension reduction, which focuses on reducing the number of variables within data to localize the necessary information [14, 15].

8.2.2.2 Deep Learning

Deep learning is a subset of ML and is based on the communication nodes present within the biological systems of humans [16]. It is important to understand the general concept of deep learning architectures and the structure of an artificial neural network (ANN). The architecture of the ANN is composed of three layers. As illustrated in Figure 8.1, the first layer is the input layer, where data is taken into the system for further processing within the succeeding layers [17]. Once data is taken into the input layer, it is transferred to the hidden layer where the artificial neurons take in a set of weight data inputs and generate a data output through completing an activation function. Lastly, the output layer takes the data and produces the given outputs.

Subsequently, deep learning serves as an extension off of a typical ANN, offering an additional two or more layers, hence gaining the term "deep" learning [18]. There are many different deep learning neural network (DNN) algorithms, with some of the most common being convolutional neural networks (CNNs), recurrent neural networks, recursive neural networks, and unsupervised pretrained networks (UPNs). The common characteristics of deep learning approaches

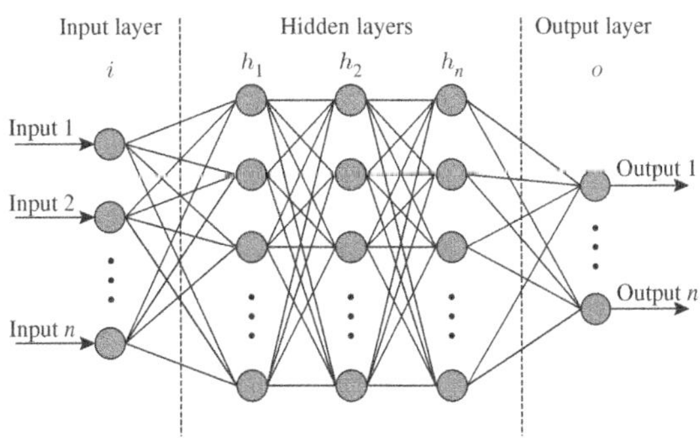

Figure 8.1 The architecture of an artificial neural network. Source: Bre et al. [17] / with permission of Elsevier.

include having strong learning ability and being extremely powerful and efficient when utilizing significantly large datasets [18]. Deep learning requires high-end graphical processing units and hardware that offer incredibly high performance to process workloads. Nevertheless, overall time requirements are larger [19]. Although these factors can be considered constraints, a higher level of accuracy is often attributed to deep learning approaches [20].

8.2.2.3 Computer Vision

Some examples of computer vision techniques can be demonstrated in subdomains such as image classification, object detection, object tracking, semantic segmentation, and instance segmentation [21]. Taking these subdomains into account, and depending on which technique is used, computer vision supports capabilities that are illustrated through software such as facial recognition, surveillance, biometrics, and self-driving cars. If developing computer system's objective was to offer speed detection, such an approach would be implemented to achieve this. One example of this approach can be shown within recent research conducted into obstacle detection and classification for high-speed autonomous driving [22]. To ensure a car is able to achieve full detection and classification of objects, the PASCAL VOC image dataset was selected, chosen as it is composed of many objects common within traffic contexts such as vehicles, pedestrians, and animals. Once the dataset was selected, it was then used to train a region-based CNN, with a Titan X GPU running on it, processing 10 frames per second (FPS). Due to such a high frame rate, the system was considered suitable for high-speed driving of autonomous vehicles, and the results from object classification and detection demonstrated full invariance when tested in different conditions such as changes in lighting or climate [22].

8.2.3 Object Recognition, Object Detection, and Object Tracking

As this research is concerned with creating a computer vision-driven prototype offering the affordance of vehicle detection, this section not just focuses on how object detection functions but also concerns itself with exploring additional surrounding computer vision approaches. These additional approaches include object recognition and object tracking. These approaches also share a dependence on one another. For instance, object detection relies on the initial establishment of the recognition of object classes to function. Similarly, object tracking relies on the accurate detection of object classes to track the object classes across video frames. Object tracking may then be employed on either previously recorded data or in real time to deliver the most up to date insights.

8.2.3.1 Object Recognition

Object recognition is a computer vision technique that enables software solutions to recognize object classes within images, video files, or livestream video sequence recordings. Once object classes have been recognized within data, they are then assigned a label that serves as a title [23]. The process of object recognition is composed of the following steps. Data is first collected from images or video files. Once obtained, features can then be extracted from the data through the use of a feature extraction algorithm. This algorithm may concern itself with acquiring certain features within each piece of data it is feed and is performed to differentiate between subsequent object classes later on in the learning process [24]. Once this process has been completed, the features that have been extracted can then be added to the selected object recognition model that will classify the features into different groupings. These groupings will then be used to help assist in analyzing newly input data and, in turn, perform recognition of the object classes that exist within the data. Overall, there are many different types of feature extraction algorithms and object classification models that work together to achieve object recognition.

8.2.3.2 Object Detection

Object detection can be understood as a computer vision technique that enables software solutions to detect and locate objects of certain classes within images, video files, or livestream video sequence recordings. Sometimes, object detection is used interchangeably with the technique of object recognition. However, there is a difference between the two. For instance, object recognition is concerned with recognizing the object within image data, whereas object detection focuses on focuses understanding what the object is but also locating its position within the given data [25]. To further make sense of how object detection functions, it is important to differentiate between some of the common terms commonly used when performing this approach. Below, object classification and object localization will be briefly examined in accordance to their role and function within object detection.

Object classification is one of the first steps necessary in completing object detection, used to predict the class within an image. This is also the core step in completing object recognition [25]. The first step of object classification is to feed a computer vision algorithm input data. Generally, the input data includes images composed of differing objects that the algorithm uses to learn and predict the objects in future data. Once the algorithm is given the input image data, it then outputs the same image data with a class label, such as one or more integers assigned to objects within the image data [26].

Once the computer vision algorithm has completed object classification, it is then able to move on to object localization. Object localization is a computer vision approach used to determine where an object is situated within an image or video file. For the algorithm to be able to correctly identify the location of a specific object, differing approaches are employed, with one of the most common being bounding box regression that is used to refine or predict localization boxes within data. Usually, the bounding box regressors are trained to identify the predefined object classes through regressing either fixed anchor boxes or region proposal networks [27]. Hence, the function of object detection serves to fulfill two main objectives: to comprehend each object class, also known as object classification, and to understand where objects are located within an image, also referred to as object localization [28].

8.2.3.3 Object Tracking

Object tracking is related to how an application can maintain the ability to stay focused on specific objects that have been detected and determining the placement of the object over various video frames. This can be employed on batched video files or in real time. The process of object tracking can be broken down into three main components. The first step includes taking a set of bounding box coordinates from object detection approaches. Once these coordinates have been obtained, an individual ID for each of the initial detections must be created. The final step is then to track each of the detected objects as they change location through each video frame while simultaneously upholding the assignment of each of the IDs [29]. Overall, the process of assigning IDs to each object allows the ability to track them across the video frames.

8.2.4 Edge Computing, Fog Computing, and Cloud Computing

When utilizing computer vision approaches, it is significant to identify the way in which data will be processed for computations and how workloads will be handled and stored. Subsequently, edge computing, fog computing, and cloud computing are architectures that offer processing and data storage capabilities. Accordingly, these architectures will be briefly considered in accordance to their roles in supporting the use of data.

8.2.4.1 Edge Computing

The first architecture may be evidenced by edge computing. The term edge computing comes about when addressing its main objective, being that it is concerned with processing data at the edge of a network [30]. Edge computing can also be defined as a distributed computing paradigm, where the management of data is completed closer to enterprise applications, such as local servers or IoT devices [31]. Taking this into account, computer vision approaches greatly rely on the speed at which data may be handled to offer real-time results. Thus, by enabling data processing at different points and allowing data to be processed wherever it is required the most, edge computing offers enterprises the ability to achieve immediate insights that can be transformed into significant processes and practices.

8.2.4.2 Fog Computing

Fog computing is another approach that can be taken to deploying computer vision applications. This, also referred to as fogging or fog networking, can be defined as a scenario where a large amount of heterogeneous, ubiquitous, and decentralized devices connect with one another to undertake processing and storage tasks [32]. These tasks can vary from supporting new services and applications, to assisting with basic networking functions. Furthermore, users that rent out these services are able to receive incentives as compensation. Thus, fog computing is incredibly similar to edge computing in the way that data is managed and processed closer to where it is originally created. Due to this, the terms edge computing and fog computing are often used interchangeably. However, one of the key differences between the two is the location of where the processing of the data takes place [32]. For instance, edge computing focuses on processing data directly on the devices close to where sensors are located, whereas fog computing ordinarily sends the computing power to a local area network (LAN). Once the data has been sent to an LAN, it is then processed either through a node, hub, router, or gateway, and the results of the processing are then sent through to the elected devices [33].

8.2.4.3 Cloud Computing

Lastly, another option may be demonstrated by cloud computing that can be defined as the on-demand delivery of computer services, including networking, software, storage, servers, and databases. The delivery of these services is offered through the Internet, also recognized by the term the "cloud" [34]. Many of these services are given acronyms, with some of the most popular being SaaS meaning software as a service, IaaS meaning infrastructure as a service, and PaaS meaning platform as a service. Furthermore, these service models are offered for remote use, allowing users utilizing applications the ability to employ them remotely and only needing to pay for as much of the service as they require [35].

8.2.5 Benefits of Computer Vision-Driven Traffic Management

This research is also concerned with understanding what benefits computer vision may offer within a traffic management context. Consequently, some examples of these benefits will be addressed below.

While the use of computer vision-driven traffic management solutions is still a relatively new concept, many benefits are now becoming noticed. One example can be illustrated by research conducted by Fedorov et al. [9], who estimated the traffic flow rate by using a video surveillance camera composed of visual data. To accomplish this research, a modified version of a faster R-CNN two-stage detector, as well as a SORT tracker, was employed. In addition, a region-based heuristic algorithm was used to sort what direction vehicles were moving. Once trials had been completed, the results demonstrated that the system was able to comprehend the number of vehicles within its field of view and understand and report on the direction of vehicle movement during heavy traffic with a less than 10% recorded mean error percentage [9].

A second benefit of using computer vision techniques within traffic management can be illustrated by its uses within real-time video processing. One example of this can be illustrated by Gulati and Srinivasan [5], who conducted research into various types of detector systems for traffic management supported by computer vision techniques. During the research, it was noted that current conventional detection systems used within traffic management include RADAR, LASER, infrared, ultrasonic, and magnetometer, all holding different competencies. However, when comparing video image processing against conventional detection systems, it was acknowledged that it held the most capabilities, including detection, presence, speed, occupancy, classification, and count of vehicles. Additional benefits include the abilities to offer real-time traffic detection, comprehensive area detection, and cost-effective quotes for installation and maintenance of traffic management solutions [5].

A third benefit of using computer vision within traffic management can be demonstrated by the way it can reduce traffic jams and congestion on busy roads through automation features. One example of this can be evidenced by research undertaken by Osman et al. [7], where a cost-effective intelligent and automated traffic control system was designed and implemented. There were two major components of the system that were detrimental to its success, including image processing and network settings. Image processing was run on a server because processing done through embedded devices would cost significantly more. Another reason that image processing was completed through a server was that it was more time efficient. With regard to network settings, a network was created in between the traffic controller and the server by utilizing HTTP protocol, alongside general network techniques. This decision is made due to their stable and well-established communication protocols for images and other data [7].

8.2.6 Challenges of Computer Vision-Driven Traffic Management

While understanding the benefits of using computer vision within traffic management, there are also technical challenges that accompany it.

8.2.6.1 Big Data Issues

Big data can be understood as information that exceeds processing, storage, and computing capacity of traditional databases and data analysis approaches [36].

Taking this into account, computer vision approaches depend on big data to fulfill their objectives. For instance, one example can be evidenced by a scenario concerned with employing an object detection algorithm for real-time detection. In this context, not only does object detection require initial training data, but also it must analyze the data presented in real time. Thus, such a reliance on big data often opens itself up to different vulnerabilities.

To prevent some of these barriers, Qiu et al. [37] undertook research into the use of ML processes for big data processing, finding that there are five critical issues that ML may be met with when handling big data. These issues include the (i) volume of data, (ii) variety of data, (iii) velocity of data, (iv) veracity of data, and, lastly, (v) the value of data. For instance, volume of data represents the quantity of data that is available. Although large scales of readily available data are often advantageous, there is also a drawback. For instance, as enterprises regularly amass terabytes to petabytes of information, the volume of big data can be considered insurmountable [38]. Because of information overload, the volume characteristic is considered one of the most significant and distinctive components of big data analysis, prompting the need for specific requirements that must be met to assist in managing it [38]. Thus, the role of computer vision approaches in handling large volumes of big data can be considered an issue not only in its management but also in the way that it must deal with the scale of differing data sources.

8.2.6.2 Privacy Issues

As evidenced above, computer vision greatly relies on big data to function. Hence, another point to consider are the precautions that must be taken to uphold and preserve privacy in certain aspects of data usage. These precautions must be taken to negate any negative consequences that may stem from poor differential privacy [39]. Overall, differential privacy can be defined as an instrument that is able to output information of an underlying dataset and withhold specific data so that it is able to serve its intended purpose. Thus, this instrument completes its objective by offering the required information, such as patterns or insights in the underlying data, but ensuring that sensitive information existing within that same dataset is not compromised [39]. Furthermore, it should be noted that differential privacy is not regarded as a binary notion, but rather a matter of accumulative risk [40]. What this means is that the privacy of an individual is not considered either breached or anonymous, but instead is given parameters that quantify data loss through different mathematical functions.

Computer vision adoption has raised privacy concerns within the facial recognition, facial expression recognition, and scene recognition domains in the recent years. This is largely due to the reason that cameras have become ubiquitous and users not being aware of their presence [41]. This reality not only calls for users to be granted the right to be aware of where cameras are situated but also acknowledges that computer vision-driven solutions must have adequate control over the real-time collection and usage of this data.

8.2.6.3 Technical Barriers

There are some current technical barriers that computer vision faces upon its implementation, such as class imbalance problems, limitation of model generalization, absence of scene understanding, and issues faced during image synthesis. Below, the significance and the true meaning of these technical issues will be considered.

One of the most common issues surrounding the implementation of computer vision approaches can be evidenced when addressing class imbalance problems. Class imbalance can be defined as a classification issue, where the distribution of instances across a given dataset is either one sided or skewed. Class imbalance problems can largely vary; however more severe instances such as where high-class imbalance occurs may call for specialized techniques and approaches to be efficient [42]. Hence, it is evident that effective classification using imbalanced data is an exceptionally important area of research due to the reason that high-class imbalance inherently exists within many real-world applications, such as cancer detection and fraud detection [43]. However, it has been noted that only a small amount of empirical research exists surrounding the use of deep learning with existing class imbalance. As a result, present research focuses its efforts on computer vision tasks with CNNs and neglects to consider the effects of big data [43].

A second technical issue to consider when implementing computer vision approaches can be evidenced by model generalization limitations. Model generalization can be defined as a model's ability to adapt properly to previously unseen or used data and draw the same distributions as the ones drawn from the one used to create the initial model with the original data [44]. Furthermore, the act of data collection, data cleaning, data preparation, and building models can all bring different challenges; however, one of the most significant of these challenges is determining how to distinguish if the model will predict future distributions [45]. Taking this into account, computer vision functions by exploiting a dataset to train and assess a model, achieved by splitting up datasets into a training set and a test set [46]. However, the downside to this approach is that the trained model then has the same data distribution across the training and test set, meaning that there is a likelihood that the data distribution will vary from the data used throughout training and testing, such as different camera angles or object scales.

8.3 Research Methodology

This section discusses how a traffic management prototype may be developed, tested, and analyzed. Sections 8.3.1, 8.3.3, 8.3.5, 8.3.6 include the research questions, the adapted design-based approach, and hardware and software instruments. The design-based research methodology is acknowledged alongside issues that may be encountered with this methodology and how they may be mitigated.

8.3.1 Research Questions and Objectives

The primary research question is as follows:

1. How can computer vision be used to support traffic flow analysis?

Accompanying questions of this research include the following:

2. What is required within the planning, development, and implementation phases of a prototype that performs accurate vehicle detection?
3. How will the performance of a vehicle-detection prototype be measured?
4. What are the barriers to the development of a vehicle-detection prototype?

As stated in Section 8.3.3, the central aim of the research is to investigate and discuss a vehicle-detection prototype. The objectives are to describe the planning, development, and implementation phases of a modern vehicle-detection prototype, to propose a modern computer vision-driven prototype that can perform accurate vehicle detection for traffic flow analysis, and to understand how to assess the performance of a prototype based on evaluation criteria.

8.3.2 Study Design

The design-based research methodology is composed of several key characteristics that make it an appropriate option for this research. A design-based research methodology can be defined by its approach to placing focus on the design and testing of a meaningful intervention [47]. In this context, a meaningful intervention may be regarded not only as the simple act of designing and testing but also a way of representing specific academic claims and welcoming the relationship between theory, design, and practice [47]. Four features must be aligned with a design-based methodology to ensure effective interventions that include learning frameworks, outlined affordances of the selected instructional tools, demonstration of domain knowledge, and contextual limitations [47]. The design-based

Table 8.1 The adaptation of the DBR methodology for this research.

Analysis of practical problem	Development and design	Iterative testing	Reflection and redesign
The first step involves the completion of research into current approaches to traffic management using computer vision. Current approaches can be viewed within the literature review	The second step involves the development and design of a prototype by taking into account previous strengths and weaknesses of the literature	The third step includes testing the prototype	The fourth step involves reflection of the testing and recalibrating if necessary

research methodology is often split up into four stages including the analysis of practical problems, development of solutions, iterative cycle testing, and reflection of design principles.

8.3.2.1 Selection Rationale

The primary motivation for selecting the design-based research methodology can be demonstrated through the objectives both the methodology and the research share in common. As the objective of the research is related to the creation of a prototype through the implementation of a phase-based approach, this complements the phases also laid out within the design-based research methodology (Table 8.1).

8.3.2.2 Potential Challenges

Upon selecting the design-based research methodology, it is essential to identify any potential challenges that may be presented when aligning it with the research to mitigate them. Below, some of the potential challenges that come with electing the design-based research methodology will be outlined.

It must be noted first that one of the most significant challenges that come with utilizing the design-based research methodology is how it operates in different research contexts. For instance, the design-based research methodology places significant emphasis on objectivity, reliability, and validity, illustrating that these features are vital in ensuring that scientifically sound research is effectively produced [48]. However, when these features are managed in a different context, such as in a controlled experimentation environment, the way they are prioritized may also vary. This in itself places the researcher in a difficult position. On the one hand, it is vital that objectivity is advocated for when facilitating meaningful interventions. However, on the other hand, the researcher may find themselves in both of the roles of advocate and critic [48].

A second challenge that may come about may be demonstrated by uncertainty arising when implementing this approach [49]. For instance, it has been posited that the efficacy of the design-based learning methodology is strongly dependent upon who conducts it. This bears a significant contrast to other approaches such as grounded theory or further dedicated experiments that commonly explain and outline their processes in particular detail. Although this flexibility gives researchers the ability to adopt a range of differing design processes that may fall under the design-based learning methodology, it is reasoned that some of these processes fail to wholly define the advanced phases of design. Similarly, an additional issue to address is that some of these design frameworks also fail to properly convey how they accompany research components [49].

Taking these challenges into account, there are a few provisions that may be taken to help mitigate them. First, this research may place significant focus on the approach to the design-based learning methodology for the research objectives and ensuring they fall in line with the defining key phases. Next, the way in which the design-based methodology complements its research counterpart may be realized through ensuring that a clear definition of its key phases is offered. Through presenting a well-defined classification of each of the phases, Easterday et al. [49] reasons that it resolves any uncertainty, informs new researchers through illustrating best practices, and, overall, strengthens communication among the research community.

8.3.3 Adapted Study Design Research Approach

Once the design-based research methodology had been identified, it was then applied to the research by adapting each stage of its phase-based approach. To convey the way the design-based research methodology was adapted to the research, each of its four phases will be outlined below with an accompanying table.

Analysis: As the first phase of the design-based research methodology is concerned with the analysis of practical problems, the research focused on analyzing present-day literature surrounding the usage of computer vision (Table 8.2).

Development: The second phase focused its efforts on prototype design and development (Table 8.3).

Table 8.2 Applying phase 1 of the DBR methodology to the prototype.

Phase 1 – Analysis	Summary of the adapted study design research approach
Analysis of practical problems by researchers and users faced when implementing solutions [50]	• Adopted an academic foundation through analysis of the use of computer vision in the present day, taking note of approaches and software/hardware instruments utilized • Completed a literature review encompassing computer vision fundamentals • Used the literature to guide the research through assessment of current day barriers to assist in effective implementation

Table 8.3 Applying phase 2 of the DBR methodology to the prototype.

Phase 2 – Development	Summary of the adapted study design research approach
Development of solutions informed by present-day design principles and technological innovations [50]	• Outlined weaknesses in current approaches to computer vision usage in traffic management • Acknowledged weaknesses through designing a prototype utilizing recent technological innovations. This was achieved through implementation of alternative instruments • Identified and adhered to relevant present-day software design principles • Obtained the instruments and developed the computer vision prototype

Testing: The third phase was concerned with testing the prototype (Table 8.4)

Redesign: The final phase of the design-based research concentrated on developing design principles that may be implemented in additional iterations (Table 8.5).

8.3.4 Selected Hardware and Software

The tools required to create the traffic management prototype were identified through examining past and current instruments implemented in other projects and looking into recent technological innovations in the domain of computer vision. Below, each of the selected hardware and software components

Table 8.4 Applying phase 3 of the DBR methodology to the prototype.

Phase 3 – Testing	Summary of the adapted study design research approach
Iterative cycles of trialing and fine-tuning of solutions in practice [50]	• Opted to implement iterative cycles of testing of the traffic management prototype • Identified and recorded the results from the testing of the prototype. Results were then used to inform suitability of design principles

Table 8.5 Applying phase 4 of the DBR methodology to the prototype.

Phase 4 – Redesign	Summary of the adapted study design research approach
Reflect upon iterations to employ additional design principles. Use the design principles to improve performance of future solutions in following iterations [50]	• Opted to implement iterative cycles of redesign to assist the testing phase of the traffic management prototype • Reflected upon the results of the prototype in each iteration • Researched and employed modern-day design principles to redesign

will be briefly examined with regard to their function and rationale behind selecting them.

8.3.4.1 Hardware: The NVIDIA Jetson Nano Developer Kit and Accompanying Items

The NVIDIA Jetson Nano Developer Kit can be understood as a small yet strong computer that enables users to run multiple neural networks in parallel and, in turn, allows the ability to harness capabilities such as image classification, segmentation, object detection, recognition, and tracking and implement them into applications in real time [51]. The rationale behind selecting the NVIDIA Jetson Nano Developer Kit over alternative modern day embedded system options is as follows. First, analyzing the current literature concerned with the use of embedded systems for computer vision in traffic management illustrates that one of the most popular choices are the range of Raspberry Pi-embedded devices [52, 53]. Due to this, there is a gap in the literature surrounding the employment of the NVIDIA Jetson range for traffic management and, in particular, the newest model NVIDIA Jetson Nano Developer Kit, released in March 2019. Additionally, further choice behind the selection of the NVIDIA Jetson Nano Developer Kit can be evidenced by price (*NVIDIA®* [54]). One final reason for the choice in the NVIDIA Jetson Nano Developer Kit can be evidenced by the objectives set out by the research, as well as analyzing the requirements necessary to design and develop the vehicle detection prototype. For instance, as the research focuses on specifically vehicle detection approaches, it requires a significant amount of processing power. Hence, composed of a Quad-core ARM Cortex -A57 MPCore central processing unit (CPU), the NVIDIA Jetson Nano Developer Kit offers a sufficient processing power suitable for modern computer vision applications [55].

To use the NVIDIA Jetson Nano Developer Kit, a microSD card, micro-USB power supply, WiFi supply, a computer display using either HDMI or DP, and

Table 8.6 The hardware components obtained and utilized for the design and development of the prototype.

Selected hardware	Technical specifications/additional information
NVIDIA Jetson Nano Developer Kit	CPU: Quad-core ARM A57 at 1.43 GHz GPU: 128-core Maxwell Memory: 4 GB 64-bit LPDDR4 25.6 GB/s Storage: uses a microSD (not inclusive) Camera: 2x IPI CSI-2 PHY lanes Connectivity: Gigabit Ethernet, M.2 Key E Display: HDMI and DP USB: 4x USB 3.0, USB 2.0 Micro-B
SanDisk microSD card: 64 GB	Speeds: up to 170 MB/s Read, up to 90 MB/s Write
HP X900 mouse	Wired USB connection
Keyboard	Wired USB connection
Computer monitor	Apple Mac 21-in. desktop
Belkin power adapter	Power supply: 5 V = 2 A
Belkin Micro-USB cable	Length: 3.3FT/1M cable
Ethernet cable	Length: 20 m
Raspberry Pi V2 camera module	Size: 25 mm x 24 mm x 9 mm Video modes: 1080p30, 720p60 and 640 x 480p60/90 Still resolution: 8 megapixels Sensor: Sony IMX219 Sensor resolution: 3280 x 2464 pixels

a USB keyboard and mouse are required. Taking this into account, it is recommended to utilize a microSD card that holds 16 gigabytes (GB) minimum [51]. Thus, a 64 GB microSD card was obtained for the prototype. Moreover, it is also recommended to use a 5v2A micro-USB power supply and so this was also obtained.

The NVIDIA Jetson Nano Developer Kit also has camera capabilities, and to utilize this feature for the development of the prototype, a Raspberry Pi V2 camera module was obtained due to its cost-effective price and its compatibility across a wide range of differing edge computing devices [56] (Table 8.6).

8.3.5 Hardware Proposed

8.3.5.1 Software Stack: NVIDIA Jetpack SDK and Accompanying Requirements (All Iterations)

The underlying software considered foundational is the NVIDIA Jetpack Software Development Kit (SDK). This is one of the most comprehensive solutions for building applications and solutions leveraging deep learning on the NVIDIA Jetson Nano Developer Kit as it is composed of many different components that may be used for design and development, including a range of software libraries and application programming interfaces (APIs) [57]. Furthermore, the NVIDIA Jetpack SDK also includes the latest operating system (OS) images for the NVIDIA Jetson range and sample applications and demos, developer tools, and documentation [57]. Each time the Jetpack SDK is updated to a newer version, so too are the versions of the libraries included within the Jetpack SDK version. Both versions of Jetpack 4.2 and 4.4 have been obtained and utilized for the purpose of design and development of the prototype by being written to two identical microSD cards to test out approaches that utilize both older and newer versions of the software (Table 8.7).

8.3.6 Software Proposed

The design-based research methodology will guide the iterations of the vehicle-detection prototype. The first iteration will take into account the phases of environment setup, design and development, implementation, analysis, and redesign of the prototype. Before carrying out the vehicle detection approaches, the vehicle detection environment needs to be prepared for design and development. To achieve setting up the environment, the following steps need to be completed. Firstly, the researcher will write the image to the microSD card by downloading the Jetson Nano Developer Kit SD card image available on

Table 8.7 Software obtained and utilized for the development and design of the prototype.

Software stack	Version	Technical specifications/additional information
Ubuntu	18.04	Linux distribution offering an open-source operating system (OS)
NVIDIA Jetpack SDK	4.2 and 4.4	SDK for building AI applications
CUDA	10 and 10.2	Parallel computing platform
CuDNN	8	Library offering deep learning frameworks
TensorFlow	N/A	Software library for dataflow and differentiable programming
TensorRT	N/A	SDK for deep learning inference
Python	2.7 and 3.6	Code used to execute models on both approaches
SSD MobileNet model	V2	Neural network used to run on the Jetson Nano
COCO dataset	N/A	Used within the first approach to train data
Open Images dataset	V6	Used for second approach to train data
Repositories from GitHub	N/A	Two repositories forked and cloned across the approaches

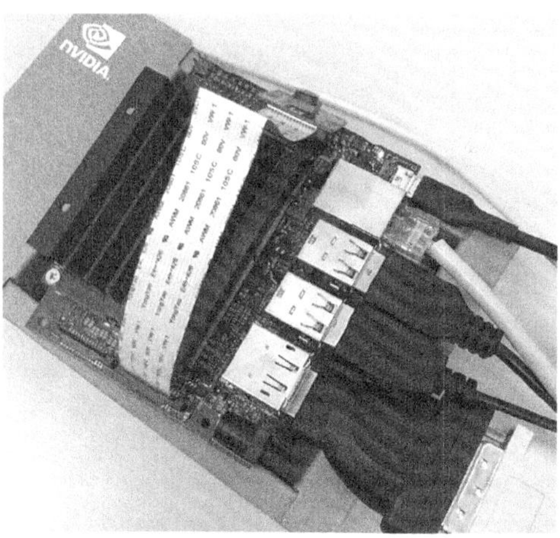

Figure 8.2 The Jetson Nano Developer Kit after setup. Source: NVIDIA.

the NVIDIA website. Secondly, the NVIDIA Jetson Nano Developer Kit needs to be setup. Thirdly, as this is the first time that the Jetson Nano Kit is powered on, it will require an initial setting up of the configurations. Chapter 9 will describe these steps in more detail and the testing and the analysis of the results (Figure 8.2).

8.4 Conclusion

With this chapter, a vehicle-detection prototype has been proposed as a way to record live traffic flow. This plan follows a design-based research methodology and has outlined only the essential features. Future proposals may also examine a way to record data and then have it transmitted through to the prototype at a later stage. Additional features to investigate may include a graphical user interface (GUI) for user-friendliness and an API to better output vehicle detection data instead of the terminal.

While this study has moved toward developing a functional prototype, future studies may look into better customizing existing prototypes for supporting smart traffic. For instance, Salonen [58] conducted research into traffic safety, in-vehicle security, and traffic management when using driverless shuttle busses. This was achieved by analyzing the passengers' own experiences, issues, and expectations from their perspective. Future testing for vehicle detection may include feedback from drivers, those evaluating new tools for traffic flow analysis, computer-vision developers, and other researchers. Evaluating these perceptions may provide insight into additional changes or improvements for future applications.

NVIDIA is working on computer vision developments that may offer more options surrounding suitable open-source code in the future. One example is NVIDIA Metropolis, an edge-to-cloud platform supported by SDKs such as Jetpack that can be used to maintain and improve smart cities through the use of artificial intelligence (AI) [59]. Overall, vehicle detection holds much potential for assisting in conducting traffic flow analysis in smart cities in the future. Based on the software and hardware components in this chapter, the implementation of the prototype for vehicle detection commenced. The implementation and testing are discussed in Chapter 9.

This research has discussed a variety of computer vision approaches and contributed to the literature by offering an up-to-date overview and outlining practical development approaches. These will also inform other researchers who are investigating similar domains using embedded devices within the IoT.

References

1 Elgendy, M. (2019). Deep learning for vision systems. https://livebook.manning.com/book/grokking-deep-learning-for-computer-vision/chapter-1/v-8/ (accessed 24 July 2021).

2 Buch, N., Velastin, S.A., and Orwell, J. (2011). A review of computer vision techniques for the analysis of urban traffic. *IEEE Transactions on Intelligent Transportation Systems* 12 (3): 920–939. https://doi.org/10.1109/TITS.2011.2119372.

3 Grauman, K. and Leibe, B. (2010). Visual object recognition. https://cs.gmu.edu/~kosecka/cs682/grauman-recognition-draft-27-01-11.pdf (accessed 24 July 2021).

4 MetroCount (2015). RoadPod. https://metrocount.com/products/roadpod-vehicle-tube-classifier/ (accessed 19 September 2020).

5 Gulati, I. and Srinivasan, R. (2019). Image processing in intelligent traffic management. *International Journal of Recent Technology and Engineering* 8 (2S4): 213–218. https://doi.org/10.35940/ijrte.B1040.0782S419.

6 Song, H., Liang, H., Li, H. et al. (2019). Vision-based vehicle detection and counting system using deep learning in highway scenes. *European Transport Research Review* 11 (1): 51. https://doi.org/10.1186/s12544-019-0390-4.

7 Osman, T., Pysche, S., Ferdous, S., and Zaman, H. (2017). Intelligent traffic management system for cross section of roads using computer vision, *IEEE Annual Computing and Communication Workshop and Conference (CCWC)* (9–11 January 2017). Las Vegas, NV, USA: IEEE.

8 Ai, C. and Tsai, Y.J. (2016). An automated sign retroreflectivity condition evaluation methodology using mobile LIDAR and computer vision. *Transportation Research Part C: Emerging Technologies* 63: 96–113. https://doi.org/10.1016/j.trc.2015.12.002.

9 Fedorov, A., Nikolskaia, K., Ivanov, S. et al. (2019). Traffic flow estimation with data from a video surveillance camera. *Journal of Big Data* 6 (1): 73. https://doi.org/10.1186/s40537-019-0234-z.

10 IBM (21 October 2019). What is machine learning?. https://www.ibm.com/topics/machine-learning (accessed 24 July 2021).

11 Panch, T., Szolovits, P., and Atun, R. (2018). Artificial intelligence, machine learning and health systems. *Journal of Global Health* 8 (2). https://doi.org/10.7189/jogh.08.020303.

12 Qolomany, B., Al-Fuqaha, A., Gupta, A. et al. (2019). Leveraging machine learning and big data for smart buildings: a comprehensive survey. *IEEE Access* 7: 90316–90356. http://arxiv.org/abs/1904.01460.

13 Abu-nimeh, S., Nappa, D., Wang, X., and Nair, S. (2007). A comparison of machine learning techniques for phishing detection. *ECrime '07: Proceedings of the Anti-Phishing Working Groups 2nd Annual ECrime Researchers Summit.* ICPS (4–5 October 2007. Pittsburgh, Pennsylvania, USA.

14 Kanungo, T., Mount, D.M., Netanyahu, N.S. et al. (2002). An efficient *k*-means clustering algorithm: analysis and implementation. *IEEE Transactions on Pattern Analysis and Machine Intelligence* 24 (7): 881–892. https://doi.org/10.1109/TPAMI.2002.1017616.

15 Nguyen, L.H. and Holmes, S. (2019). Ten quick tips for effective dimensionality reduction. *PLoS Computational Biology* 15 (6): e1006907. https://doi.org/10.1371/journal.pcbi.1006907.

16 Amazon Web Services, Inc. (2020). Deep learning on AWS. https://aws.amazon.com/deep-learning/ (accessed 26 May 2020).

17 Bre, F., Gimenez, J., and Victor, F. (2017). Prediction of wind pressure coefficients on building surfaces using artificial neural networks. *Energy and Buildings* 158: 1429–1441. https://doi.org/10.1016/j.enbuild.2017.11.045.

18 Dargan, S., Kumar, M., Ayyagari, M.R., and Kumar, G. (2019). A survey of deep learning and its applications: a new paradigm to machine learning. *Archives of Computational Methods in Engineering*. https://doi.org/10.1007/s11831-019-09344-w.

19 Ceron, R. (05 December 2019). AI, machine learning and deep learning: what's the difference? IBM IT Infrastructure Blog. https://www.ibm.com/blogs/systems/ai-machine-learning-and-deep-learning-whats-the-difference/ (accessed 24 July 2021).

20 Kowsari, K., Brown, D., Heidarysafa, M., et al. (2017). Hierarchical deep learning for text classification. https://www.researchgate.net/publication/319968747_HDLTex_Hierarchical_Deep_Learning_for_Text_Classification (accessed 24 July 2021).

21 Liu, L., Ouyang, W., Wang, X. et al. (2020). Deep learning for generic object detection: a survey. *International Journal of Computer Vision* 128 (2): 261–318. https://doi.org/10.1007/s11263-019-01247-4.

22 Prabhakar, G., Kailath, B., Natarajan, S., & Kumar, R. (2017). Obstacle detection and classification using deep learning for tracking in high-speed autonomous driving. *2017 IEEE Region 10 Symposium (TENSYMP) (14–16 July 2017)*. Cochin, India: IEEE.

23 Bansal, M., Kumar, M., and Kumar, M. (2020). 2D object recognition techniques: state-of-the-art work. *Archives of Computational Methods in Engineering* https://doi.org/10.1007/s11831-020-09409-1.

24 Mathworks (2020). Object recognition. https://www.mathworks.com/solutions/image-video-processing/object-recognition.html (accessed 4 October 2020).

25 Vipul, J. (15 February 2018). Image recognition vs object detection: the difference. Hackernoon. https://hackernoon.com/micro-learnings-image-classification-vs-object-detection-the-difference-77110b592343 (accessed 24 July 2021).

26 Brownlee, J. (21 May 2019). A gentle introduction to object recognition with deep learning. *Machine Learning Mastery*. https://machinelearningmastery.com/object-recognition-with-deep-learning/.

27 Lee, S., Kwak, S., and Cho, M. (2019). Universal bounding box regression and its applications. https://arxiv.org/abs/1904.06805?fbclid=IwAR08OT5j2Jmi8xRksjGE7gKdIXtYhsiBz2w8I-erlmrb2sdCImbQPuUGWhw.

28 Zhao, Z.-Q., Zheng, P., Xu, S., and Wu, X. (2019). Object detection with deep learning: a review. *IEEE Transactions on Neural Networks and Learning Systems* 30: 1–21.

29 Rosebrock, A. (23 July 2018). Simple object tracking with OpenCV. PyImageSearch. https://www.pyimagesearch.com/2018/07/23/simple-object-tracking-with-opencv/ (accessed 24 July 2021).

30 Huang, D. and Wu, H. (2018). Edge clouds: pushing the boundary of mobile clouds. In: *Mobile Cloud Computing* (eds. D. Huang and H. Wu), 153–176. Morgan Kaufmann. https://doi.org/10.1016/B978-0-12-809641-3.00008-9.

31 IBM (2020). What Is edge computing. https://www.ibm.com/cloud/what-is-edge-computing (accessed 18 August 2020).

32 Naha, R., Garg, S., Georgakopoulos, D. et al. (2018). Fog computing: survey of trends, architectures, requirements, and research directions. *IEEE Access* 6: 47980–48009. https://www.researchgate.net/publication/326171764_Fog_Computing_Survey_of_Trends_Architectures_Requirements_and_Research_Directions.

33 Klonoff, D.C. (2017). Fog computing and edge computing architectures for processing data from diabetes devices connected to the medical Internet of

Things. *Journal of Diabetes Science and Technology* 11 (4): 647–652. https://doi.org/10.1177/1932296817717007.
34 Microsoft Azure (2020). What is cloud computing?. https://azure.microsoft.com/en-us/overview/what-is-cloud-computing/ (accessed 18 August 2020).
35 Wilson, A. (1 April 2018). Cloud computing: the next frontier for vision software. Vision Systems Design. https://www.vision-systems.com/factory/article/16739360/cloud-computing-the-next-frontier-for-vision-software (accessed 24 July 2021).
36 Najafabadi, M., Villanustre, F., Khoshgoftaar, T. et al. (2015). Deep learning applications and challenges in big data analytics. *Journal of Big Data* 2. https://doi.org/10.1186/s40537-014-0007-7.
37 Qiu, J., Wu, Q., Ding, G. et al. (2016). A survey of machine learning for big data processing. *EURASIP Journal on Advances in Signal Processing* 2016 (1): 67. https://doi.org/10.1186/s13634-016-0355-x.
38 Hadi, H.J., Shnain, A.H., Hadishaheed, S., and Ahmad, A.H. (2015). Big data and the five V's characteristics. *International Journal of Advances in Electronics and Computer Science* 2 (1): 9.
39 Ji, Z., Lipton, Z., and Elkan, C. (2014). Differential privacy and machine learning: a survey and review. https://www.researchgate.net/publication/269997816_Differential_Privacy_and_Machine_Learning_a_Survey_and_Review (accessed 24 July 2021).
40 Harvard University (2020). Differential privacy. https://privacytools.seas.harvard.edu/differential-privacy (accessed 31 July 2020).
41 Das, A., Degeling, M., Wang, X. et al. (2017). Assisting users in a world full of cameras: a privacy-aware infrastructure for computer vision applications. In: *2017 IEEE Conference on Computer Vision and Pattern Recognition Workshops (CVPRW)*, 1387–1396. IEEE. https://doi.org/10.1109/CVPRW.2017.181.
42 Brownlee, J. (22 December 2019). A gentle introduction to imbalanced classification. *Machine Learning Mastery*. https://machinelearningmastery.com/what-is-imbalanced-classification/.
43 Johnson, J.M. and Khoshgoftaar, T.M. (2019). Survey on deep learning with class imbalance. *Journal of Big Data* 6 (1): 27. https://doi.org/10.1186/s40537-019-0192-5.
44 Google Developers (2020). Generalisation. https://developers.google.com/machine-learning/crash-course/generalization/video-lecture (3 August 2020).

45 Bateman, B. (2018). Challenges of generalisation in machine learning. https://blogs.oracle.com/datascience/challenges-of-generalization-in-machine-learning (accessed 24 July 2021).

46 Dai, J. and Lin, S. (2018). Image recognition: current challenges and emerging opportunities. Microsoft Research. https://www.microsoft.com/en-us/research/lab/microsoft-research-asia/articles/image-recognition-current-challenges-and-emerging-opportunities/ (accessed 24 July 2021).

47 Anderson, T. and Shattuck, J. (2012). Design-based research. *Educational Researcher* 41: 16–25. https://doi.org/10.3102/0013189X11428813.

48 The Design-Based Research Collective (2003). Design-based research: an emerging paradigm for educational inquiry. *Educational Researcher* 32 (1): 5–8. https://doi.org/10.3102/0013189X032001005.

49 Easterday, M., Rees Lewis, D., and Gerber, E. (2014). Design-based research process: problems, phases, and applications. In: *Proceedings of International Conference of the Learning Sciences*, vol. 1 (eds. B. Penuel, S. Jurow and K. O'Connor), 317–324. ICLS.

50 Hong, J., Cochrane, T., and Withell, A. (27 November 2018). Developing a design-based research methodology for designing MR technologies for mountain safety. ASCILITE 2018, Deakin University, Geelong, Australia. https://www.researchgate.net/publication/329174708_Developing_a_design-based_research_methodology_for_designing_MR_technologies_for_mountain_safety.

51 Nvidia Developer (5 March 2019). Getting started with the jetson nano developer kit. https://developer.nvidia.com/embedded/learn/get-started-jetson-nano-devkit (accessed 24 uly 2021).

52 Bhusari, S., Patil, S., and Kalbhor, M. (2015). Traffic control system using Raspberry Pi. *Global Journal of Advanced Engineering Technologies* 4 (4): 3.

53 Lokesh, S. and Reddy, P. (2014). An adaptive traffic control system using Raspberry PI. Internet of Things for Traffic Control. https://www.researchgate.net/publication/282612189_An_Adaptive_Traffic_Control_System_Using_Raspberry_PI (accessed 24 July 2021).

54 Jetson nano developer kit. (n.d.). Nvidia Developer. https://developer.nvidia.com/embedded/jetson-nano-developer-kit.

55 Nvidia Developer (2020). Hardware for every situation. https://developer.nvidia.com/embedded/develop/hardware (accessed 25 August 2020).

56 Raspberry Pi (2020). Buy a camera module V2 – Raspberry Pi. https://www.raspberrypi.org/products/camera-module-v2/ (accessed 4 September 2020).

57 Nvidia Developer (2020). JetPack [Nvidia]. http://docs.nvidia.com/jetson/jetpack/introduction/index.html (accessed 4 September 2020).

58 Salonen, A.O. (2018). Passenger's subjective traffic safety, in-vehicle security and emergency management in the driverless shuttle bus in Finland. *Transport Policy* 61: 106–110. https://doi.org/10.1016/j.tranpol.2017.10.011.

59 Nvidia (2020). Build smarter cities through AI. https://www.nvidia.com/en-us/industries/smart-cities/ (accessed 8 November 2020).

9

Implementation and Evaluation of Computer Vision Prototype for Vehicle Detection

Gabrielle Bakker-Reynolds, Emre Erturk, Istvan Lengyel, and Noor Alani

Eastern Institute of Technology, School of Computing, Hawke's Bay, New Zealand

9.1 Prototype Setup

9.1.1 Introduction

Vehicle detection is a useful environment understanding process for describing two-dimensional (2D) and three-dimensional (3D) scenes in computer vision. It is frequently used for the development of computer vision modules for autonomous cars and robotics. Robust vehicle detection and scene understanding are key tasks for visual sensors (cameras) in autonomous cars for being able to interpret and act within a dynamic environment [1, 2]. Cameras play a significant role in autonomous driving. They are capable of providing rich information including object detection and distances of objects given in traffic scenes. Hence, developing a prototype that is capable of handling large amounts of image data has been a challenging task especially where the prototypes are designed from locally available resources.

Subsequently, this chapter explores how vehicle detection can be applied to support traffic flow analysis with the aim to design and develop a computer vision prototype from available resources during 2020. This chapter also assumes a design-based approach, composed of two iteration-based approaches employed to develop a prototype for vehicle detection utilizing a nano-computer. The first iteration was the phases of environment setup and choice of methods for most iterations. Next were the design and development, implementation, analysis, and redesign of the prototype. The approaches are analyzed according to the accuracy, processing time, cost, and overall suitability.

Cyberphysical Smart Cities Infrastructures: Optimal Operation and Intelligent Decision Making, First Edition. Edited by M. Hadi Amini and Miadreza Shafie-khah.

9.1.2 Environment Setup

Before testing the different detection approaches, the vehicle detection environment was prepared for design and development. To achieve this, the following steps were completed.

Step 1. Writing the Jetson Nano OS image to the MicroSD card: The first step was to write the image to the microSD card, completed by downloading the Jetson Nano Developer Kit SD card image available on the NVIDIA website (*Getting Started With* [3]). Following the download of the image, a graphical program titled Etcher, which allows users to flash operating system (OS) images to SD card or USB drives, was utilized. To achieve this, Etcher was firstly downloaded and installed onto the computer at the computer lab room at EIT. Once installed, the Etcher application was launched, the option "select image" was selected, and the zipped Jetson Nano Developer Kit SD card image file was selected. Once selected, the microSD card was inserted into a USB-C micro/SD card reader, and the reader was connected to the computer. The Etcher application then prompted the flash command that was selected as shown below. Once the image was written, the SD card was then ejected via the Files application, and the microSD card was removed from the computer.

Step 2. Setup: Following, the NVIDIA Jetson Nano Developer Kit was setup. To achieve this, the developer kit was unpackaged and placed on top of the developer kit box. The microSD card with the newly written OS image was inserted into the slot located underneath the Jetson Nano module. Next, a USB keyboard and mouse were connected, alongside connecting to the computer monitor via an HDMI cable. For livestreaming capabilities, the Raspberry Pi V2 camera module was also connected to the Jetson Nano. Lastly, the Jetson Nano was connected via a 5V2A Micro USB power supply, prompting the Jetson Nano to power on and boot automatically. A green LED light lit up upon startup indicating that it had powered on correctly. See below for a photo of the complete Jetson Nano Developer Kit module (see Figure 9.1).

Step 3: First boot: As this was the first time that the Jetson Nano Kit was powered on, it required an initial setting up of the configurations. This was demonstrated by the Jetson Nano prompting review and acceptance of the NVIDIA Jetson software end-user license agreement (EULA) and requests for the user to select their system language, layout of keyboard, and time zone. Once preferences were set, username and password were required to be created, and the final step was to log in to the system.

Step 4: Choice of analysis: Once the Jetson Nano Kit was set up, the way the collected data would be analyzed was posited. After researching into various methods of analysis, the formula below was used to calculate the averages for both the accuracy and processing time of the images, MP4 file, and livestreaming

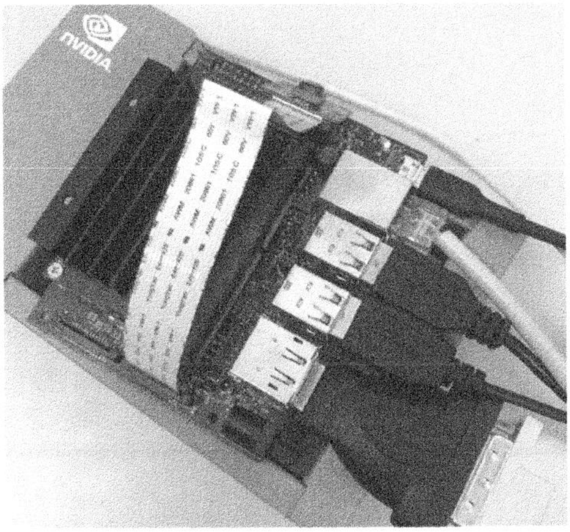

Figure 9.1 The Jetson Nano Developer Kit after setup. Source: NVIDIA.

for all iterations excluding the last iteration outlined later in this chapter. This formula was chosen as it calculates the accuracy of the model with a 95% confidence interval. The formula used to calculate the accuracy of the models (using the CPU Confidence intervals) is

$$\overline{X} \pm (t \times s) / \sqrt{n} \tag{9.1}$$

where n is the number of observations and s and t are the standard deviation and t-test value, respectively.

9.2 Testing

9.2.1 Design and Development: The Default Model and the First Iteration

The decision was then made to clone and install the repository titled "jetson-inference" located on GitHub [4]. For the first iteration of this approach, the offered object detection demo was tested out. This demo object detector utilizes a deep neural network approach called DetectNet, and the default model used is the single-shot multibox detector (SSD) MobileNet v2 trained on the common objects in context (COCO) dataset. To use DetectNet, the project was built from source by adhering to the instructions located within the repository. Subsequently, the following steps were taken. First, both Git and CMake were installed on the Jetson Nano. Following, cloning of the repository was completed, and the directory was changed to point toward the repository. A build directory was then created, and CMake was run within the build directory to configure it. Once these commands were input into the terminal, it prompted a downloader tool to appear within the terminal interface. Only SSD MobileNet was downloaded from this. After the model was selected, a downloader tool titled "Pytorch Installer" appeared. The downloader tool can be used to install the software library Pytorch onto the Jetson Nano. Pytorch's primary function within the jetson-inference repository is to enable transfer learning to retrain neural networks. As transfer learning is looked into in the subsequent iteration, Pytorch was installed. Furthermore, as the Jetson Nano was running Jetpack 4.4, the Python 3.6 version of Pytorch 1.1.0 was installed. The final step was to compile the project build. To achieve this, it was critical to ensure that the directory was still located within the correct directory within the terminal window. Once the project was cloned, installed, and configured, it was then possible to commence testing of the sample object detection model.

9.2.2 Testing (Multiple Images)

The first test concerned itself with assessing the performance of the model on multiple images of vehicles. To achieve this, a sequence of vehicle images was processed, completed by launching the DetectNet model with the path to the directory containing the vehicle image data. Moreover, it was important to test on a significant number of images to assess whether there was consistency in the results. Subsequently, the model was run on 100 images of vehicles taken from the Open Images V6 dataset. More information about this dataset is provided later in this chapter. To test the model on the vehicle data, three parameters were set within the command, including specifying the model, the image input, and where to save the image results.

9.2.3 Analysis (Multiple Images)

Accuracy: When launching the command, the image results were saved to the specified directory. At the same time, additional information was output via the terminal, including the class identification number, bounding box coordinates, and confidence values. Upon a check of the output directory, each one of the vehicle images appeared to offer a colored bounding box overlaying the image and accompanying confidence values. To recognize the average, the first 25 images of the 100 images were taken and input into a table using Microsoft Excel. Some of the image data offered more than one datapoint, making it a challenge to calculate an average across the sample size. Due to this, images that included a singular vehicle were counted. Taking this into account, additional 12 vehicle images were included to make up a sample size of 25, and if any of the images included more than one vehicle, they were skipped. Once the datapoints were input in Microsoft Excel, the average and a 95% confidence interval were calculated, giving the average of 84.65 ± 0.05.

Timing: The model was evaluated in relation to processing time. Assessment was completed by evaluating the timing report output via the terminal upon image processing. As the terminal outputs a timing report for each image that it processes, a sample size of the first 25 images was taken from the 100 images. The analysis took into account the total time it took for the CPU and compute unified device architecture (CUDA) to process each image. Once the datapoints were input in Microsoft Excel, the average and a 95% confidence interval were calculated for both CPU and CUDA processing. For CPU, the average processing time was 49.33 ± 5.65%. For CUDA, the average processing time was 55.85 ± 5.28%. Table 9.1 illustrates the datapoints obtained, alongside the averages and the averages using the formula.

The costs of the current approaches were calculated by looking at the costs associated with traffic counters created by the business MetroCount, as these are the products being used currently within the Hawke's Bay region to conduct traffic flow analysis. Furthermore, to conduct traffic flow analysis using the pneumatic tubes, specifically for using the RoadPod VT4 product, MetroCount states that two road survey field kits are required [5]. Each road survey field kit includes road nails, road cleats, road rubber flaps, and rubber road tubes. Although the RoadPod VT4 is a standard product created by MetroCount for traffic analysis, there is no public information detailing its cost. As people are employed to count traffic within the Hawke's Bay region, this was also factored in under the current approaches taken to traffic flow analysis. It was estimated that individuals are paid the New Zealand minimum wage to carry out this task. A list of the components is shown in Table 9.2.

9.2.4 Testing (MP4 File)

An open-source MP4 file of vehicles was also run on the model. The stock photos and video website Pexels were searched to find a suitable video file to fit this purpose. One video file was selected for use titled "daylight traffic on a camera angled tilt road." This video was composed of 24 seconds of footage of different vehicles travelling across a road [6]. Once selected, the video file was then downloaded off of the Pexels website and saved locally under the "jetson-inference" directory. The terminal was then launched and commands were input.

Analysis (MP4 File)

Accuracy: This was assessed by evaluating the output confidence values associated with the vehicle detections. To calculate an average, eight datapoints were taken from the output generated via the terminal and input into Microsoft Excel.

Table 9.1 Datapoints obtained alongside the average and average using the formula.

Approach 1, Iteration 1: default model – images			
Image	**Confidence value (%)**	**CPU processing time (ms)**	**CUDA processing time (ms)**
1	74.90	114.9	114.5
2	51.80	49.57	64.5
3	80.30	46.6	50.39
4	88.60	46.5	51
5	85.60	46.8	54.1
6	89.50	47.2	54.6
7	88.00	46.9	55.3
8	94.40	47.9	55.3
9	68.40	46.2	51.3
10	98.60	45.7	56.2
11	87.40	46	54.5
12	98.70	46.3	52.8
13	58.70	45.8	46.7
14	91.10	46.4	55.2
15	86.30	46.8	56
16	62.00	46.1	46.2
17	87.30	47.2	54.4
18	96.60	46.5	53.2
19	95.00	45.7	50.9
20	97.30	46.6	51.2
21	95.30	46.3	52.1
22	80.50	47.1	60.3
23	96.30	45.5	52.7
24	92.90	46.5	52.6
25	70.80	46.2	50.2
Average:	**84.65**	**49.3308**	**55.8476**
Average using formula:	**84.65 ± 0.05**	**49.33 ± 5.65**	**55.85 ± 5.28**

Table 9.2 Comparison of the prototype and current approaches.

The vehicle detector for traffic flow		The current traffic flow approach	
NVIDIA Jetson Nano Developer Kit	**$169.75**	**12 × 70 mm road nails (×2)**	**See below**
Raspberry Pi V2 camera module	$42.60	10 × Fig. 8 road cleats (×2)	
Ethernet cable	$15.00	2 × road rubber flaps (×2)	
60GB microSD card	$60.00	1 × 30 m natural rubber road tube (×2)	$324.50
Micro USB power supply cable	$20.00	Employees	$18.80 P/H
Micro USB power supply wall charger	$20.00		
HDMI cable	$20.00		
Mouse	$10.00		
Keyboard	$20.00		
Monitor	$200.00		
Total cost:	**$327.25**	**Total cost:**	**$649.00**

The first datapoint output via the model was considerably higher; however, this was to be expected when starting up the model. Due to this, this datapoint was considered an outlier and so was made exempt when calculating the average. Once the datapoints were input within Microsoft Excel, the average and 95% confidence interval were calculated, giving the average of 84.79 ± 0.0821.

Timing: The performance of the model was assessed by evaluating 14 datapoints offered via the terminal. Seven datapoints were concerned with model processing timing for CPU, whereas the other seven offered model processing timing for CUDA. To remain consistent with the accuracy assessment, the first datapoint was made exempt when calculating the average timing for the model as it was higher than all of the following datapoints. Once the datapoints were input in Microsoft Excel, the average and 95% confidence interval were calculated for both CPU and CUDA processing. For CPU, the average processing time was 56.3 ± 2.24 ms. For CUDA, the average processing time was 47.21 ± 2.36 ms. Table 9.3 illustrates the datapoints obtained, alongside the averages and the averages using the formula.

Table 9.3 Datapoints obtained alongside the average and average using the formula.

Approach 1, Iteration 1: default model – MP4 file			
Image	**Confidence value (%)**	**CPU processing time (ms)**	**CUDA processing time (ms)**
1	88.50	54.9	46.4
2	68.20	56.2	45.7
3	84.70	54.3	45.4
4	76.80	55.2	46.2
5	96.60	54.1	45.1
6	86.30	59.9	51.5
7	92.40	59.5	50.2
Average	**84.79**	**56.3**	**47.21**
Analysis using formula	**84.79 ± 0.0821**	**56.3 ± 2.24**	**47.21 ± 2.36**

9.2.5 Testing (Livestream Camera)

The final test analyzed the performance of the model in relation to its livestream capabilities for real-time vehicle detection. To achieve this, the Jetson Nano was setup within an outdoor environment. There were no changes in the outdoor setup than in the previous iterations; however an Ethernet cable that offered a longer cord alongside an extension lead was utilized. The Jetson Nano was then tested on a singular car, moving at 5 km/h. An image illustrating the Jetson Nano after setting up can be evidenced in Figure 9.2.

The Analysis (Livestream Camera)

Accuracy: As the camera was pointed toward the vehicle, it started to pick up detections. Initially, the camera was not angled straight and sat on a slight slope, detecting the moving vehicle as incorrect miscellaneous objects. Once the camera position was corrected by being pointed directly at the camera, it then appeared to register the vehicle as a car, outputting this data via the terminal. With the model operated by outputting data every few frames, and as the vehicle moved closer to the camera, the confidence values increased. To calculate the average across the livestreaming of the vehicle, the first 10 datapoints were taken from when the start of where the model began to register the object class correctly. Once the datapoints were input within Microsoft Excel, the average and 95% confidence interval were then calculated, giving the average of 96.97 ± 0.0462%.

Figure 9.2 Livestream setup of the Jetson Nano.

Timing: Accompanying each output was a timing report, offering the average timing in milliseconds across CPU and CUDA real-time processing. The associated timing datapoints along with the accuracy datapoints were obtained and input within a table in Microsoft Excel. Table 9.4 offers the data obtained and used to calculate the timing for both the CPU and CUDA. Once the datapoints were input in Microsoft Excel, the average and 95% confidence interval were calculated for both CPU and CUDA processing. For CPU, the average processing time was 52.29 ± 0.638 ms. For CUDA, the average processing time was 50.03 ± 1.14 ms. Table 9.4 illustrates the datapoints obtained, alongside the averages and the averages using the formula.

Redesign: Although this iteration yielded sufficient results for vehicle detection, it was posited whether utilizing the method of transfer learning could heighten accuracy or optimize time efficiency even more so. To ensure a fair comparison, the following iteration would make use of the existing model, be tested using the same images and MP4 file, and include the same livestream setup.

9.3 Iteration 2: Transfer Learning Model

9.3.1 Design and Development

The SSD MobileNet was then retrained with Pytorch through using vehicle images selected from the Open Images V6 database via transfer learning. This approach does not require the need for the model to be trained from the beginning, making it very time efficient [7]. As this section focuses on the design and development

Table 9.4 Datapoints obtained alongside the average and average using the formula.

Approach 1, Iteration 1: default model – livestream camera			
Image	**Confidence value (%)**	**CPU processing time (ms)**	**CUDA processing time (ms)**
1	96	51.9	50.1
2	97.1	52.4	46.57
3	96.4	52.7	50.4
4	96.6	53.8	52.3
5	96.3	51	48.9
6	96.8	52.1	49.5
7	97.7	53.5	51.7
8	97.8	51.7	49.7
9	97.7	52.5	51
10	97.3	51.3	50.1
Average	**96.97**	**52.29**	**50.027**
Analysis using formula	**96.97 $\pm$ 0.0462**	**52.29 $\pm$ 0.638**	**50.03 $\pm$ 1.14**

of the prototype, the SSD MobileNet and the Open Images V6 dataset will both be examined below briefly.

Combining the SSD-300 SSD with the support of a MobileNet backbone, the SSD MobileNet object detector is a standard neural network architecture designed to be used for real-time object detection [7]. First revealed in a 2017 paper at Cornell University, MobileNets can be understood as convolutional neural networks specifically designed for use on mobile and embedded devices [8]. They are based on a streamlined architecture that use depth-wise divisible convolutions to create lightweight neural networks (see Figure 9.3). Furthermore, MobileNets employ two hyper parameters that introduce a balance between latency and accuracy of results [8]. The SSD-MobileNet base model used for transfer learning was pre-trained on the PASCAL visual objects class (VOC) dataset, a significantly large dataset used as a benchmark in object category recognition and detection.

The Open Images dataset is a dataset composed of over nine million images [9]. Each of these images has been annotated with corresponding image level labels and offers additional features such as bounding boxes and segmentation masks for each object, as well as mapping out visual relationships. Furthermore, the Open Images V6 dataset contains 16 million bounding boxes produced for 600 object classes trained on 1.9 million images. This makes the Open Images V6 dataset the biggest present-day dataset offering object location annotations [9]. Moreover,

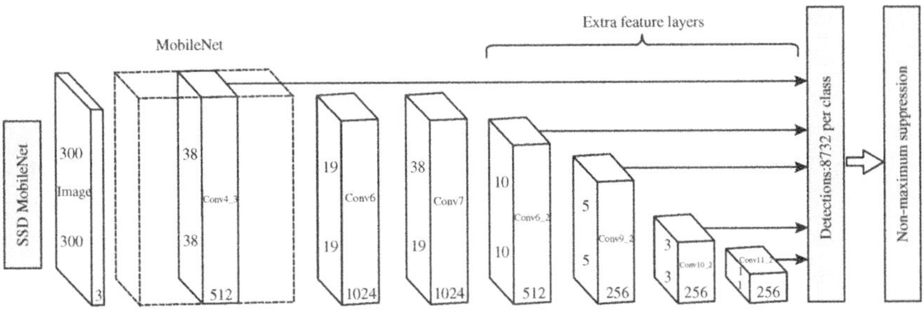

Figure 9.3 Architecture of the SSD MobileNet object detector model [7].

it should also be noted that the bounding boxes around each object have been manually drawn by professional annotators to ensure high levels of accuracy and reliability when using the dataset. The primary strength in mapping out visual relationships is that it supports visual relationship detection, a developing task involving structured reasoning [10].

To retrain the SSD MobileNet, a few steps had to be completed first. In order, these steps included setting up the Jetson Nano, downloading the necessary data, limiting the amount of data used, training the SSD MobileNet model, and converting the SSD MobileNet model to open neural network exchange (ONNX) format. Below, each of these steps will be discussed.

Firstly, to set up the Jetson Nano, the Pytorch training code available within the previously downloaded repository was utilized. However, before the code could be used, the base model SSD MobileNet was obtained, and additional python packages were installed [4]. Following, data from the Open Images V6 dataset was downloaded. However, since the Open Images V6 dataset contains large quantities of data, it was important to limit the amount of data so that there was sufficient space on the Jetson Nano. Time efficiency was also taken into account, and so this iteration concerned itself with using a small amount of data to retrain the model. Consequently, there was the option to run a Linux command titled "stats only" specifying the required object classes prior to download of the images. This option also showed the number of images available for download under each class before downloading them.

Prior to selecting the object classes, the list of object classes available for download was viewed on the Open Images V6 dataset website. Upon viewing the list, five object classes were selected for training, including "car," "vehicle," "vehicle registration plate," "truck," and "motorcycle." Once downloading the object class statistics surrounding these object classes, it showed that there was a total of 116 806 number of images available for download, alongside a total bounding box count of 317 790.

The total number of images were then divided into three categories including training image data, validation image data, and testing image data, with additional information given on the number of images of object classes that were available from the total categories. Once discovering that there was a total of 93 854 images available to train the model, the number of images selected from this number for download was minimized to 2500. To download the specified number of images, the syntax for the command remained the same, but the "stats only" parameter was replaced with the "max images" parameter. This was imperative because without detailing the maximum number of images required for download, all of the images would have been downloaded by default that could overload the Jetson Nano. Once the specified number of images were downloaded, the SSD MobileNet model was able to commence training. To achieve this, the python script was run using the appropriate command argument options. The full list of options that could be chosen from can be evidenced in Table 9.5.

The batch size hyperparameter command identifies the number of samples that are processed through the neural network prior to the model being updated [11]. Conversely, the epoch hyperparameter command specifies the number of complete passes through the image training data [12]. For this iteration, the number of batch sizes was set to the number 4, and the number of epochs was set to 80 to ensure training did not use up all available memory or take a long period of time. Once the commands were input, the Jetson Nano was left to complete training. Once training was completed, the final step was to convert the model to ONNX format. To convert the model, it was important to ensure that the directory was still located within the "SSD" subdirectory of the jetson-inference repository. Once the model accomplished conversion to ONNX format, the development phase of the model was completed.

Table 9.5 List of command options for training the SSD MobileNet model [48].

Argument	Default	Description
–data	data/	The location of the dataset
–model-dir	models/	Directory to output the trained model checkpoints
–resume	None	Path to an existing checkpoint to resume training from
–batch-size	4	Try increasing depending on available memory
–epochs	30	Up to 100 is desirable but will increase training time
–workers	2	Number of data loader threads (0 = disable multithreading)

9.3.2 Test (Multiple Images)

Once the design and development phase were completed, the model was then tested via the terminal. Once the commands were input, it prompted a pop-up box of the image results to appear in consecutive order, alongside output results via the terminal.

9.3.3 Analysis (Multiple Images)

Accuracy: Upon assessing the results of the images, a manual look illustrated that the model was able to detect the vehicles in the images. A total of 100 images were run on the model. From this, a sample size of 25 was taken from the total amount to provide the average confidence value across the results. The rationale behind calculating an average confidence value was to understand and compare the performance of the model against the previously tested working default model. Similar to the above iteration, it was planned for the data to be input into Microsoft Excel, multiplied, and then divided by the total sample size. However, the results provided images that included more than one detection within the image data, making it difficult to calculate an average. Therefore, images with more than one confidence value of vehicles were excluded to provide an accurate average. To make up for the excluded data, additional five vehicle images were included to maintain the sample size, counting on from 25. If the image included more than once confidence value, it was skipped. Once the datapoints were input in Microsoft Excel, the average and 95% confidence interval were calculated. This gave an average of 91.19 $\pm$ 0.043%.

Timing: Corresponding to the above working iteration, an assessment of model timing was completed by evaluating the timing report output via the terminal upon image processing. The sample size of the first 25 images was taken from the 100 images, taking into account the total time it took for the CPU and CUDA to process each image. Once the datapoints were input in Microsoft Excel, the average and 95% confidence interval were calculated for both CPU and CUDA processing. For CPU, the average processing time was 30.608 $\pm$ 1.64%. For CUDA, the average processing time was 38.068 $\pm$ 2.02%. Table 9.6 illustrates the datapoints obtained, alongside the averages and the averages using the formula.

9.3.4 Test (MP4 File)

The same MP4 file from Pexels utilized in the previous iteration was used for testing. To commence testing, the terminal was launched, and the directory was

Table 9.6 Datapoints obtained alongside the average and average using the formula.

Approach 2, Iteration 1: transfer learning model – images			
Image	**Confidence value (%)**	**CPU processing time (ms)**	**CUDA processing time (ms)**
1	98.50	46.9	54.7
2	93.80	35.9	48.6
3	65.10	33.5	38.6
4	99.70	34.6	41.2
5	96.90	33.4	40.2
6	90.00	28.7	35.4
7	90.80	29.1	39.6
8	83.40	28.8	35.4
9	99.30	28.7	35.3
10	98.60	29.1	35.7
11	86.30	29.2	37.5
12	68.80	29.2	36.2
13	95.00	28.4	29.2
14	82.60	29.5	39.3
15	97.70	28.7	36.1
16	93.30	30	37.8
17	87.60	28	36.6
18	65.80	29.9	37
19	97.10	29	34.3
20	98.80	30.3	36.5
21	98.70	28.7	40.6
22	99.30	29.2	37.1
23	93.50	29	36.1
24	99.60	28.4	33.4
25	99.50	29	39.3
Average	**91.19**	**30.608**	**38.068**
Average using formula	**91.19 ± 0.043**	**30.608 ± 1.64**	**38.068 ± 2.02**

pointed toward the jetson-inference directory. Upon testing, the MP4 file opened up and started automatically playing. When vehicles appeared within the MP4 file, bounding box and confidence value for each vehicle were present.

9.3.5 Analysis (MP4 File)

Accuracy: The model was assessed by evaluating the output confidence values associated with the vehicle detections to determine accuracy. Eight confidence values were taken from the output generated via the terminal and input into Microsoft Excel to calculate an average. Similar to the previous iteration, the confidence value output via the model was considerably higher; however, this was to be expected when starting up the model. Owing to this, the datapoint was considered an outlier and was not counted when calculating the average. Once the datapoints were input in Microsoft Excel, the average and 95% confidence interval were calculated, giving an average of 88.91 ± 0.043%.

Timing: 14 datapoints were then assessed to offer an average of the timing across the model. Seven datapoints were concerned with model processing timing for the CPU, and the other seven were concerned with offering model processing timing for CUDA. The first datapoint was made exempt when calculating the average timing for the model as it was higher than all of the following datapoints. Once the datapoints were input in Microsoft Excel, the average and 95% confidence interval were calculated for both the CPU and CUDA processing. For CPU, the average processing time was 44.6 ± 9.80 ms. For CUDA, the average processing time was 33.4 ± 7.88 ms. Table 9.7 illustrates the datapoints obtained, alongside the averages and the averages using the formula.

9.3.6 Test (Livestream Camera)

The final test evaluated the model within a real-time context. To ensure lightening and weather conditions remained consistent, this test was also was recorded on the same day as the test carried out for the first iteration.

9.3.7 Analysis (Livestream Camera)

Accuracy: This test was a reproduction of the one carried out for the first iteration with the default model, with the only exception being that the code employed to test out the model within the terminal differed. Consequently, the vehicle was driven across the camera frame in the same manner, and the camera was angled in the same position. Once the datapoints were input within Microsoft Excel, the average and 95% confidence interval were then calculated, giving the average of 99.17 ± 0.243%.

Table 9.7 Datapoints obtained alongside the average and average using the formula.

Approach 1, Iteration 1: transfer learning model – MP4 file			
Image	Confidence value (%)	CPU processing time (ms)	CUDA processing time (ms)
1	89.00	68	52.2
2	86.00	45.6	34.5
3	92.40	38.6	27.9
4	89.00	40.5	30.4
5	88.70	39.2	29.3
6	88.80	40.5	30.4
7	88.50	39.2	29.3
Average	**88.91**	**44.51429**	**33.2857**
Analysis using formula	**88.91 ± 0.043**	**44.6 ± 9.80**	**33.4 ± 7.88**

Timing: The average time it took to detect the model was calculated by obtaining data from timing report output via the terminal. Once the datapoints were input in Microsoft Excel, the average and 95% confidence interval were calculated for both the CPU and CUDA processing. For CPU, the average processing time was 35.278 ± 1.065 ms. For CUDA, the average processing time was 33.869 ± 1.041 ms. Table 9.8 illustrates the datapoints obtained, alongside the average and the averages using the formula.

9.3.8 Redesign

Upon the assessment of the development and analysis of the previous working iterations, the research found two improvements for the following iteration. The first improvement has to do with the sample sizes in previous iterations. For instance, earlier in the first iteration, only seven datapoints were used to calculate an average. However, utilizing sample sizes composed of only seven datapoints may limit the accuracy evaluation of the vehicle detections. Consequently, an iteration utilizing a wider range of datapoints may give a more accurate reading of the true accuracy performance of the detection models. Furthermore, the second improvement can be made when addressing how accuracy of the prototype is analyzed. Presently, as demonstrated in the previous iterations, accuracy of the prototype has been analyzed by calculating an average using Microsoft Excel. However, this method does not account for variables such as true positive or false positive vehicle detections. Furthermore, if this information is obtained, it may offer additional

Table 9.8 Datapoints obtained alongside the average and average using the formula.

Approach 1, Iteration 1: transfer learning model – livestream camera.			
Image	Confidence value (%)	CPU processing time (ms)	CUDA processing time (ms)
1	99.3	35.8	34.8
2	99.4	37.38	35.59
3	99.5	34.3	34.1
4	99.1	33.3	34.1
5	99.2	37.3	35.1
6	99.2	36.7	35.5
7	98.3	33.6	31.4
8	99.3	34.6	32.3
9	99	35.5	33.4
10	99.4	34.3	32.4
Average	**99.17**	**35.278**	**33.869**
Analysis using formula	**99.17 ± 0.243**	**35.278 ± 1.065**	**33.869 ± 1.041**

insights into the performance of the model, also offering a measure accuracy based upon these additional variables.

9.4 Iteration 3: Increased Sample Size and Change of Accuracy Analysis (Images)

9.4.1 Design and Development

The fourth iteration was concerned with implementing the improvements outlined earlier in this chapter. As the first improvement, focus was placed on retrieving a sample size of 50 datapoints of images for both the default model and the transfer learning model, doubling the sample sizes employed in the previous iterations that focus on using images. The second improvement implemented a confusion matrix on the collected datapoints. A confusion matrix is a table with four different combinations of true and predicted values and was chosen to show the relationship between the observed and predicted datapoints during the performance of the vehicle detections [13]. It is used to assist in getting additional insights. This iteration will be concerned with testing the same image data from both the default model and the transfer learning model.

9.4.2 Testing

The research obtained images of both the default model and the transfer learning model. The same steps as in the first iteration were used, but with a few changes. First, this iteration obtained a larger sample size. Second, because the testing image data remained the same as the image data outlined in the previous iterations, this iteration obtained image data for its sample size by utilizing images numbered from "50" onward within the output folders, contrasting the previous iterations that took image data counting from the first image. This approach to data collection ensured that the data samples for both the default model and the transfer learning model were still tested on the same images; however, the data samples taken differed from the previous ones. Third, the change can be seen from the way each of the 50 datapoints for the default and the transfer learning model was input into Microsoft Excel. Instead of skipping images without a vehicle object class, the image was still counted to assess whether or not the model could correctly identify this. Furthermore, if either of the models detected a vehicle within an image that did not contain a vehicle, this was also counted. Once the data was generated by the models, it was input into Microsoft Excel and arranged into two columns, with one column giving the true value and the other giving the models' predicted value.

9.4.3 Analysis

A confusion matrix was created for both the default model and the transfer learning model's generated output datapoints to perform analysis. The confusion matrix for each of the two models was calculated by manually assigning the obtained datapoints generated into one of four values. These four values included "true positive" (TP), "false positive" (FP), "false negative" (FN), and "true negative" (TN). TP was employed when the predicted value was the same as the true value, FP was used when the model predicted a vehicle that was non-existent in the true images, FN was used when the model did not predict a vehicle in an image that contained a vehicle, and TN was used when the model did not predict a vehicle in an image that did not contain a vehicle. Table 9.9 demonstrate the two confusion matrixes for both the default model and the transfer learning model using their respective output datapoints.

9.4.3.1 Confusion Matrices

Table 9.9 *illustrates* that the default model shows the 39 datapoints that were TP, 3 that were FN, 0 that were FP, and 8 that were TN. Correspondingly, Table 9.10 demonstrates that the performance of the transfer learning model yielded different results, showing that 39 datapoints were TP, 3 that were FN, 7 that were FP, and 1 that was TN.

Table 9.9 Confusion matrix for the output of the default model.

Predicted values	Actual values
TP = 39	**FP = 7**
FN = 3	TN = 1

Table 9.10 Confusion matrix for the output of the transfer learning model.

Predicted values	Actual values
TP = 39	**FP = 0**
FN = 3	TN = 8

9.4.3.2 Precision, Recall, and F-score

Based on the information provided through the two confusion matrixes, additional insights may be gained through employing metrics that determine the precision, recall, and F-score of the two models. Firstly, precision can be understood as a way to determine the proportion of positive predictions of a model that are truly correct [14]. As in the formulas below, the precision of a model is calculated by adding the TP and FP rates and dividing the TP by that:

$$\text{Precision} = \frac{\text{TP}}{\text{TP} + \text{FP}} \tag{9.2}$$

Subsequently, precision was calculated for both the default model and the transfer learning model using the above formula. The default model offered a precision of 1.0, and the transfer learning model offered a precision of 0.84. Following, recall of the two models was also calculated to determine the proportion of actual positives predicted correctly by a model [14]. As demonstrated below, the recall of a model is calculated by adding the TP and FN rates and dividing the TP by the sum:

$$\text{Recall} = \frac{\text{TP}}{\text{TP} + \text{FN}} \tag{9.3}$$

Recall was calculated for both the default model and the transfer learning model using the above formula. Both the default model and the transfer learning model offered a recall of 0.93. The last metric calculated was the F-1 score, calculated to interpret the weighted average of both the calculated precision and recall [15]. As demonstrated below, the F-1 score can be calculated using the following formula:

$$F_1 = \frac{2}{\frac{1}{\text{recall}} \times \frac{1}{\text{precision}}} = 2 \times \frac{\text{precision} \times \text{recall}}{\text{precision} + \text{recall}} \tag{9.4}$$

The F_1-score was calculated for both the default model and the transfer learning model using the above formula. The default model offered an F_1-score of 0.96, and the transfer learning model offered an $F1$-score of 0.88.

9.5 Findings and Discussion

The general findings of the research are presented here, examining the results from testing the three media inputs. Overall, the performance of vehicle detection produced different results depending on the approach implemented. For the first approach, vehicle detection was unable to be performed as it was found that object detection implements required up-to-date software across all packages and dependencies to effectively function.

Following this line of rationale, the second approach reinforced that employing open-source code available from GitHub enabled general object detection [16]; however accuracy and speed of the prototype were compromised as the software was not optimized specifically for vehicle detection. Taking this into account, the research found that by employing transfer learning on an already pretrained model, this enabled the prototype to be trained on additional vehicle data for optimization.

The second iteration focused on the increasing batch size and epochs, yielding positive results when implementing the model for testing after it had been trained. Overall, it was revealed that this approach increased the prototypes' detection accuracy and decreased detection time across vehicle detections for all three media inputs tested. The final iteration looked into increasing the sample size and changing to different analysis methods, showing that the default model offered a better precision and $F1$-score.

Below, the technical specifications of the elected hardware and software utilized within the research will be reiterated. Following, the two working iterations, composed of the default model and the final transfer learning model, will be discussed in relation to their findings. Lastly, the findings will be discussed in relation to the research questions (as stated in the previous chapter), the suitability of the prototype for traffic flow analysis will be assessed, and possible improvements for future iterations will be discussed.

9.5.1 Findings: Vehicle Detection Across Multiple Images

The default model yielded sufficient results. The data was input into Microsoft Excel, and the average function was used to calculate an average. This same process was completed for all subsequent calculations involving averages across all media inputs in the following tests.

Findings illustrated that on average, the prototype using the default model offered an average 84.65 $\pm$0.05% across testing of images of vehicles. To accompany this, the average processing time using the Quad-core ARM A57 CPU was 49.33 ± 5.65 ms, and the average processing time offered by CUDA, using a 128-core Maxell GPU, was 55.85 ± 5.28 ms. Furthermore, it should be noted all subsequent iterations discussed below utilized the same technical specifications for the approaches as outlined in Chapter 8.

During the second iteration, the transfer learning model offered better results. Overall, the transfer learning model demonstrated an increase in average, with the prototype offering an average of 91.19 ± 0.043% across images during testing. Similarly, the average timing of processing images also decreased. For the CPU, the average processing time was 30.608 ± 1.64 ms. For CUDA, the average processing time was 38.068 ± 2.02 ms.

Taking the above findings into account, the data was subjected to three paired sample t-tests performed using Microsoft Excel to determine whether there was a significant statistical difference between the two approaches. The first t-test offered a *p*-value of 0.093. As the *p*-value was not below 0.05, this demonstrated that there was no significant statistical difference in accuracy detection between the two models. A second *t*-test was performed to address whether there was a significant difference in timing processing for the CPU, offering a *p*-value of 3.87, and a third *t*-test was performed for CUDA, offering a *p*-value of 1.55. Overall, the *t*-tests showed that there was no statistical difference in timing processing between the two approaches.

Although there was no significant statistical difference between the two models in relation to performance across accuracy and timing, the transfer learning model appeared to offer an increase in average when detecting vehicles by 6.54%. The transfer learning model also offered a decrease in timing across both CPU and CUDA processing time.

To measure the relative timing performance of the default model and transfer learning model, the speedup ratio was calculated by dividing the average processing time for the default model by the transfer learning model across both CPU and CUDA [47]. For CPU, 49.33 ± 5.65 ms produced by the default model was divided by the 30.608 ± 1.64 ms produced by the CPU for the transfer learning model. Overall, results showed that the CPU of the transfer learning model provided a 160% speedup over the default model. For CUDA, 55.85 ± 5.28 ms produced by the default model was divided by the 38.068 ± 2.02 ms produced by CUDA for the transfer learning model. Overall, the CUDA of the transfer learning model provided a 147% speed improvement over the default model.

9.5.2 Findings: Vehicle Detection Performance on an MP4 File

As demonstrated earlier in this chapter, the default model prototype offered an average of 84.79 ± 0.0821% across testing of images of vehicles. Furthermore, the average processing time using the CPU was 56.3 ± 2.24 ms, and the average processing time using CUDA was 47.21 ± 2.36 ms.

As evidenced above in the second iteration, the transfer learning model offered improved results when testing on the MP4 file, offering an average of 88.91 ± 0.002%. Similarly, the average timing of processing images also decreased. For CPU, the average processing time was 44.6 ± 9.80 ms. For CUDA, the average processing time was 33.4 ± 7.88 ms.

Taking the above findings into account, the data was subjected to three paired sample *t*-tests performed using Microsoft Excel to determine whether there was a significant difference between the two approaches. The first *t*-test offered a *p*-value of 0.027, showing no statistical difference in accuracy between the two models. A second *t*-test was performed to address whether there was a significant difference in timing processing for the CPU, offering a *p*-value of 0.034, and a third *t*-test was performed to assess CUDA, offering a *p*-value of 0.007. These two t-tests illustrate that there is a significant difference in processing timing between the two models. Furthermore, given that the averages of the CPU and CUDA offered faster processing time for the transfer learning model, the *t*-tests reinforce that the transfer learning model performed faster than the default model when processing vehicle detections on an MP4 file.

To measure the relative timing performance of the default model and transfer learning model, the speedup ratio was calculated by dividing the average processing time for the default model by the transfer learning model across both CPU and CUDA [17]. For CPU, 56.3 ± 2.24 ms produced by the default model was divided by the 44.6 ± 9.80 ms produced by the CPU for the transfer learning model. Overall, results showed that the CPU of the transfer learning model provided a 126% speedup over the default model. For CUDA, 47.21 ± 2.36 ms produced by the default model was divided by the 33.4 ± 7.88 ms produced by CUDA for the transfer learning model. Overall, results showed that CUDA of the transfer learning model provided a 140% speedup over the default model.

9.5.3 Findings: Vehicle Detection on Livestream Camera

As demonstrated earlier in this chapter, the default model prototype offered an average 96.97 ± 0.0462 across testing of the camera livestreaming in real time. Furthermore, the average processing time for CPU was 52.29 ± 0.638 ms. For CUDA, the average processing time was 50.03 ± 1.14 ms.

In the second iteration, the transfer learning model offered better results on the livestream camera. This is evidenced by an offered average 99.17 ± 0.243% across testing of livestream camera. Timing was also decreased. For instance, for the CPU, the average processing time was 35.278 ± 1.065 ms. For CUDA, the average processing time was 33.869 ± 1.041 ms.

Taking the above findings into account, the data was subjected to three paired sample *t*-tests performed using Microsoft Excel to determine whether there was a significant difference between the two approaches. The first t-test offered a *p*-value of 1.86, showing no statistical difference in accuracy between the two models. A second *t*-test was performed to address whether there was a significant difference in timing processing for the CPU, offering a *p*-value of 1.14, and a third *t*-test was performed for CUDA, offering a *p*-value of 1.42.

Although there was no significant statistical difference between the models in relation to performance of accuracy and timing, it should be noted that the transfer learning model still offered an increase in average when detecting the vehicle by 2.2%. Furthermore, the transfer learning model also offered a decrease in timing across both CPU and CUDA processing time. To measure the relative timing performance of the default model and transfer learning model, the speedup ratio was calculated by dividing the average processing time for the default model by the transfer learning model across both CPU and CUDA (*Predicting and Measuring Parallel Performance*, 2012). For CPU, 52.29 ± 0.638 ms produced by the default model was divided by the 35.278 ± 1.065 ms produced by the CPU for the transfer learning model. Overall, results showed that the transfer learning model offered speedup of 140%. For CUDA, 50.03 ± 1.14 ms produced by the default model was divided by the 33.869 ± 1.041 ms produced by CUDA for the transfer learning model. Overall, results showed that the CUDA of the transfer learning model also provided a speedup of 140% over the default model.

9.5.4 Findings: Iteration 3

As demonstrated above, the final iteration was employed to assess whether additional changes, such as an increase in sample size, and differing analysis methods would offer further insights. To achieve this, the default model and the transfer learning model were tested on 50 images, doubling the sample sizes obtained in the previous iterations concerned with image processing. Furthermore, the two models were assessed in relation to precision, recall, and *F*1-scores. The default model offered a precision of 1.0, a recall of 0.93, and an F_1-score of 0.96. Conversely, the transfer learning model offered a precision of 0.84, a recall of 0.93, and an F_1-score of 0.88. Overall, this illustrates that the default model performed better than the transfer learning model in relation to precision and F_1-scores. However, it should be noted that the transfer learning model performed better in relation to average calculated in previous iterations.

Upon assessing the results of the images within the "trafficmodel2" folder, a manual look illustrated that the model was able to detect the vehicles in the images. This is shown in Figure 9.4.

A total of 100 images were run on the model. From this, a sample size of 25 was taken from the total amount to provide the average confidence value across the results. The rationale behind calculating an average confidence value was to

Figure 9.4 Quantitative result analysis using Open Images dataset (V6). The green overlay bounding boxes represents the vehicle detection accuracy using the proposed prototype. Source: Used with the permission of Microsoft.

understand and compare the performance of the model against the previously tested working default model.

9.5.5 Addressing the Research Questions

The research questions were as follows:

(1) How can computer vision be used to support traffic flow analysis?
(2) What is required within the planning, development, and implementation phases of a prototype that performs accurate vehicle detection?
(3) How will the performance of a vehicle-detection prototype be measured?
(4) What are the barriers to the development of a vehicle-detection prototype?

Addressing the first question, one of the most critical requirements to ensure an effective prototype creation was the employment of a research methodology that aligned with the research's questions and objectives. For this, the adoption of the design-based research methodology served as a suitable approach as it underlined the importance of iteration-based research and offered a phase-based framework that could be referred to during the planning, development, implementation, and redesign lifecycle. Adherence to the framework provided the research with a structure and ensured that appropriate steps were taken in each phase. Furthermore, performing iterations allowed for the research to learn from previous iterations and, in doing so, improve the prototype.

Additionally, the ability to troubleshoot issues or errors during the testing phases of the prototype was made apparent. It was found that to effectively troubleshoot issues, reliance was often placed on websites that such as GitHub or Stack Overflow as they are very community focused, offering platforms that enable developers to share information surrounding software-related issues.

Another requirement that proved to be an important component was undergoing enough research into the selection of hardware and software used to develop the prototype. Initially, the hardware for the prototype was selected prior to the choice of the software stack, which was a straightforward process. This was because once the nano-computer model was selected, the instructions and the accompanying tools were available from the manufacturer's website.

To address the second research question, evaluation criteria were posited to measure the performance of the prototype and were implemented in each analysis section of each iteration. As the prototype focuses on vehicle detection, the most critical evaluation benchmark was ensuring a high level of accuracy when detecting vehicles. Furthermore, to understand whether the prototype held the capabilities of real-time vehicle detection, the speed at which processing vehicles occurred was another assessment marker set in place. Finally, general assessment of the prototypes' suitability for traffic flow analysis is discussed further below. Also, finally, there were a few different barriers encountered when developing the vehicle detection prototype. The two most prevalent challenges were to do with cost and time. These two factors may have had an impact on the final results of the research. For example, to keep costs down, free and open-source tools were used when possible.

A final barrier can be addressed by many of the incompatibilities faced when opting to use the Jetson Nano for prototype development. The Jetson Nano is considered a relatively new embedded device. Due to this, the research often faced restrictions underpinning the development of the prototype. Some of the most prevalent restrictions included the lack of working open-source code and significant conflicts with software popular with similar embedded devices. However, as there is a further increase in development on the Jetson Nano, it is posited that these issues will be improved in the future.

9.5.6 Assessment of Suitability

High accuracy was the most important efficacy marker of the prototype. Taking this into account, it was found that it was crucial that the computer vision model was optimized for vehicles to ensure there was no chance of false positives. Consequently, this line of rationale endorses the transfer learning models' iteration outlined in the section on design and development, as it also offered the highest rate of accuracy. However, the final iteration concerned itself with analyzing both of the two models in relation to precision, recall, and corresponding $F1$-scores to generate further insights. Employing these metrics illustrated that the default model performed better than the transfer learning model in this regard.

The prototype did not appear to be hindered by processing timing when tested on vehicle images. The MP4 file and real-time livestreaming seem to serve as potentially suitable modes of vehicle detection processing for traffic flow analysis. As the prototype was optimized to detect only vehicles across six object classes, this speeded up processing time across all three media inputs. However, though timing of vehicle detection does not appear to pose as an issue, the terminal does not automatically add up the total of vehicles detected, but rather displays them one after the other in order of sequence across any media input, which is not user-friendly. Future directions for this are discussed in the section on future improvements.

Furthermore, it was found that during the design and development phase of the research, an intermediate level of knowledge was required for successful implementation. This included knowledge surrounding setting up the device and sourcing and implementing compatible software and open-source code to run it. However, once the prototype was set up, it was a straightforward process to begin running it on the different media inputs. Due to this, it may serve as a suitable option or capturing traffic flow once the initial setup of the prototype has been completed.

The prototype also proved to be a cost-effective option and appeared to rival costs associated with current approaches implemented for traffic flow analysis. The use of the prototype may extend local governmental use, providing an accessible way for enterprises to seek insights into traffic flow for their business context. Furthermore, the prototype may be suitable from a commercial standpoint because it upholds the privacy of vehicles and the vehicle owners.

Also, the prototype may also be viewed as beneficial from a sustainability standpoint due to its low power consumption. This is significant to note, as the prototype utilizes the Jetson Nano as its embedded device to run the models on; the technical specifications for its small size still boast 128 CUDA cores, 4GB RAM, and Quad-core ARM Cortex A57 processor [18]. However, when it comes to energy required to run, there is no trade-off, as the Jetson Nano only requires 5–10 W of power consumption [19].

Another affordance of the prototype is that it offers users a great level of flexibility pertaining to media inputs. For instance, typically, a general use case of the prototype would oversee the utilization of its livestreaming camera capabilities to capture real-time traffic flow. However, a secondary option presents MP4 recordings. In this way, the prototype may capture traffic flow by recording the vehicles via a camera and then running the recording through the prototype at a later time.

9.5.7 Future Improvements

Based on the assessment of suitability, improvements may be put forth for future iterations. The first recommendation can be evidenced by the choice of hardware, as some of the hardware tools were cost-effective options. Due to this, future improvements may look to employ hardware that may offer better results. One example of this may be the Raspberry Pi V2 module. Although the Raspberry Pi V2 module serves as an effective instrument in testing of real-time livestreaming capabilities, there are cameras compatible with the Jetson Nano that offer a higher picture resolution.

Another option to consider for future iterations may be utilizing a wireless WiFi option. In this research, Ethernet cables were employed; however, they may not always be regarded as the most suitable option when looking to employ the prototype within a commercial context. Future improvements may focus on utilizing tools that support wireless capabilities on the nano-computer. Taking advantage of wireless capabilities presents many additional benefits, such as increasing flexibility of where the devices can be placed, also making the overall setup of the prototype more secure. Furthermore, this would also allow for the monitor attached to the device to be located somewhere remotely.

A final hardware addition may be protective gear for the devices, especially when being subject to weather conditions. This research did not take into account physical durability of the prototype; however, if the prototype is to be implemented within a real-life traffic flow context, it is imperative that it is properly shielded by weather conditions. Subsequently, employing some form of protective-like apparatus may aid the prototype in this way.

Future improvements may also look at incorporating a graphical user interface (GUI) in replacement of the terminal. Implementing a GUI may not only increase user-friendliness but also encourage the employment and overall exploration of the prototype within a local government or enterprise context. Moreover, the development of a GUI tailored toward the prototype could also be involved in automating the process of vehicle detection counting using an application programming interface (API). Overall, these options hold the potential to remove the need in a scroll through the output data reporting on vehicle detections, saving time, and increasing overall usability.

9.6 Conclusion

Overall, the findings of the research illustrate that the prototype may be used as a potential tool in the future to aid traffic analysis. Once the prototype is running, it has demonstrated that it holds the capabilities of detecting vehicles automatically. One of the current modes of traffic counting is by employing people to count vehicles manually. Implementing a tool such as the prototype would enable a higher level of automation and not place as much reliance on people to carry out these tasks. In addition, the technology used for the prototype is more cost-effective than the current tools used within the local region. Overall, it is important to have greater cloud and IoT adoption regionally [20] in order for smart transportation to be implemented on a large scale.

Within this study, a vehicle detection prototype was developed, tested, and evaluated to serve as a way to focus on the general functionality of computer vision prototype built from locally available resources. The results of the final iteration utilized a transfer learning-based computer vision model that shows great potential for the locally developed prototype. Further improvements in hardware are recommended before this can be implemented in a real context. These improvements include camera upgrades to offer better resolution for livestream capture, wireless connection to give more flexibility, and protective apparatus to shield the prototype from weather conditions. Alternatively, as suggested by Loresco [21], for intelligent traffic light operation using computer vision with an Android system, traffic data was transmitted via a CCTV to a Raspberry Pi 3 microcontroller.

Another direction for the future may be changing the code used to run the prototype. The Jetson Nano is a relatively new embedded device, and some of the open-source code will continue to be updated and improved. One possible tool is NVIDIA Metropolis, an edge-to-cloud platform supported by SDKs such as Jetpack that can be used to maintain and improve smart cities through the use of artificial intelligence (AI) [22]. Overall, the vehicle detection prototype holds much potential for assisting in conducting traffic flow analysis in the future. Consequently, the findings of this research may inform future investigations into similar domains using embedded devices.

References

1 Ai, C. and Tsai, Y.J. (2016). An automated sign retroreflectivity condition evaluation methodology using mobile LIDAR and computer vision. *Transportation Research Part C: Emerging Technologies* 63: 96–113. https://doi.org/10.1016/j.trc.2015.12.002.

2 Alani, N.H.S. (2019). Improved stixels towards efficient traffic-scene representations. Doctoral thesis. Auckland, New Zealand: Auckland University of Technology. https://openrepository.aut.ac.nz/handle/10292/12545.

3 Nvidia Developer (2019). *Jetson nano developer kit*. https://developer.nvidia.com/embedded/jetson-nano-developer-kit (accessed 24 July 2021).

4 Franklin, D. (2020). *Dusty-nv/jetson-inference* [C++]. https://github.com/dusty-nv/jetson-inference (accessed 24 July 2021).

5 MetroCount (n.d.). *RoadPod*. https://metrocount.com/products/roadpod-vehicle-tube-classifier/ (accessed 24 July 2021).

6 Pexels (n.d.). *Daylight traffic on a camera angled tilt road: free stock video.* https://www.pexels.com/search/videos/cars%20driving/ (accessed 24 July 2021).

7 Franklin, D (2020). *jetson-inference.* GitHub. https://github.com/dusty-nv/jetson-inference/blob/master/docs/pytorch-ssd.md.

8 Howard, A.G., Zhu, M., Chen, B. et al. (2017). MobileNets: efficient convolutional neural networks for mobile vision applications. *ArXiv:1704.04861 [Cs].* http://arxiv.org/abs/1704.04861.

9 Google (2020). *Open images V6.* https://storage.googleapis.com/openimages/web/factsfigures.html (accessed 24 July 2021).

10 Kuznetsova, A., Rom, H., Alldrin, N. et al. (2020). The open images dataset V4: unified image classification, object detection, and visual relationship detection at scale. *International Journal of Computer Vision* 128 (7): 1956–1981. https://doi.org/10.1007/s11263-020-01316-z.

11 Kandel, I. and Castelli, M. (2020). The effect of batch size on the generalisability of the convolutional neural networks on a histopathology dataset. *ICT Express* https://doi.org/10.1016/j.icte.2020.04.010 6: 312–315.

12 Georgevici, A.I. and Terblanche, M. (2019). Neural networks and deep learning: a brief introduction. *Intensive Care Medicine* 45 (5): 712–714. https://doi.org/10.1007/s00134-019-05537-w.

13 Nisbet, R., Miner, G., and Yale, K. (2018). Model evaluation and enhancement. In: *Handbook of Statistical Analysis and Data Mining Applications*, 2e (eds. R. Nisbet, G. Miner and K. Yale), 215–233. Academic Press. https://doi.org/10.1016/B978-0-12-416632-5.00011-6.

14 Google Developers (n.d.). *Classification: precision and recall.* https://developers.google.com/machine-learning/crash-course/classification/precision-and-recall (accessed 24 July 2021).

15 Wood, T. (n.d.). *F-Score.* DeepAI. https://deepai.org/machine-learning-glossary-and-terms/f-score.

16 Bakker, G. (2020). Gbakkel/jetson-inference. https://github.com/gbakkel/jetson-inference (accessed 24 July 2021).

17 Gabb, H., Corden, M., Rosenquist, T. et al. (2012). *Predicting and measuring parallel performance.* Intel. https://software.intel.com/content/www/us/en/develop/articles/predicting-and-measuring-parallel-performance.html?language=en.

18 Nvidia Developer (n.d.). *Hardware for every situation.* https://developer.nvidia.com/embedded/develop/hardware (accessed 24 July 2021).

19 E Linux (n.d.). *Jetson nano.* https://elinux.org/Jetson_Nano (accessed 24 July 2021).

20 Erturk, E. (2017). An incremental model for cloud adoption: based on a study of regional organizations. *TEM Journal* 6 (4): 868–876.

21 Loresco, P. (2018). Intelligent traffic light system using computer vision with android monitoring and control. *TENCON 2018-2018 IEEE Region 10 Conference*, pp. 2461–2466). IEEE. https://doi.org/10.1109/TENCON.2018.8650084.

22 Nvidia (n.d.). *Build smarter cities through AI*. https://www.nvidia.com/en-us/industries/smart-cities/ (accessed 24 July 2021).

10

A Review on Applications of the Standard Series IEC 61850 in Smart Grid Applications

Youcef Himri[1,6], S. M. Muyeen[2], Farhan Hameed Malik[3], Saliha Himri[1], Khairol Amali bin Ahmad[4], Nachida Kasbadji Merzouk[5], and Mustapha Merzouk[6]

[1] *Université Tahri Mohamed Béchar, Faculté des Sciences Exactes, B.P 417 route kenadsa, Béchar 08000, Algeria*
[2] *Curtin University, School of Electrical Engineering, Computing and Mathematical Sciences, Perth, WA 6102, Australia*
[3] *Institute of Applied Technology, Abu Dhabi Polytechnic, P.O. Box 111499, Abu Dhabi, United Arab Emirates*
[4] *Universiti Pertahanan Nasional Malaysia, Faculty of Engineering, Department of Electrical and Electronics Engineering, Kem Sungai Besi, 57000 Kuala Lumpur, Malaysia*
[5] *Unité de Développement des Equipements Solaires/UDES, Centre de Développement des Energies Renouvelables/CDER, Route Nationale N° 11, BP 386, Bou-Ismail, 42415 Tipaza, Algeria*
[6] *Université Saad Dahlab Blida, Laboratoire FUNDamental and Apply Physics Laboratory (FUNDAPL), Algeria*

10.1 Introduction

Useful information and informantion exchange mechanisms have always played an important role in the functioning of the energy system in real time. Initially, the telephone was utilized to communicate the line loads to the control center, to dispatch operators, and to perform switching operations at substations [1]. In the mid-1930s, the telephone line was used to control the switching-based control units remotely and to provide status and control over the contacts' position [2]. As digital communications became a viable option in the 1960s, the data acquisition systems were installed to collect substation measurement data automatically [3–5]. Therefore, the control of electrical networks for many years is done using computers, and some examples of these have been applied to the control/ automation of substations in the early 1970s [6].

The first applications were absolutely in the field of bulk power transmission, as a natural extension of the trend to centralized control rooms for such systems. During the last decade, more or less, advances in computing power and software

Cyberphysical Smart Cities Infrastructures: Optimal Operation and Intelligent Decision Making, First Edition. Edited by M. Hadi Amini and Miadreza Shafie-khah.

have led to the application of computers to the control/automation of substations on a much broader basis [6], as the technology advancements are happening continuously. These developments are accelerating the modernization of the energy system with intelligent electronic devices (IEDs) that help to control and monitor immensely the energy system. The intelligent network could do energy management, security, and reliability enhancements with the new communication technology of the IEC 61850. Furthermore, IEC 61850 has grown beyond the substation boundaries to cover substations to substations, substations to control centers, asset condition monitoring, and synchrophasor applications, as well as transmissions in other areas of the power grid [7].

The objective of this chapter is to present a detailed technical review and outline the applications of IEC 61850 in the smart grid. First, we briefly introduce the definition of smart grid followed by a discussion about the basics of IEC 61850 architecture model and its functional hierarchy and IEC 61850 class model, and also the features and roles of IEC 61850 are described in details. Finally, we discuss the relevant applications and advantages of IEC 61850. At the end, conclusion and perspective of this study are given.

10.2 Overview of IEC 61850 Standards

Nowadays, smart grid structure has played a crucial role. This term was invented in 2005 by Amin and Wollenberg in their work entitled "Towards a Smart grid" [8]. The authors defined the smart grid as an electric power grid using computer and communication technology to realize a network that can monitor and heal itself. Moreover, there are five definitions, among many others, from Federal Energy Regulatory Commission (FERC), IEEE, Department of Energy (DOE), General Electric (GE), and Wikipedia in [9]. All these have the same purpose, which is increasing the reliability of electric power delivery, improving the efficiency, integrating the renewable energy, and reducing the cost of electric energy to consumers [10–12].

The smart grid uses widely IEC 61850 as a common communication standard in the power system (PS). It is a popular standard in the smart grid domain for substation automation and distribution systems. Figure 10.1 shows an example of the IEC 61850 protocol. This was designed based on the Utility Communications Architecture (UCA), an open communication standard for electricity companies defined by the electric power research institute in the early 1990s. Originally, IEC 61850 was developed for communications in substations. Subsequently, it expands the concepts and principles outlined in the original version beyond the substation fence [13].

IEC 61850 "Communication networks and systems in substations" is a standard published in 2003 by the International Electrotechnical Commission (IEC) for power utilities capable of providing and ensuring the integration and interoperability between IEDs that perform such functions as protection, control,

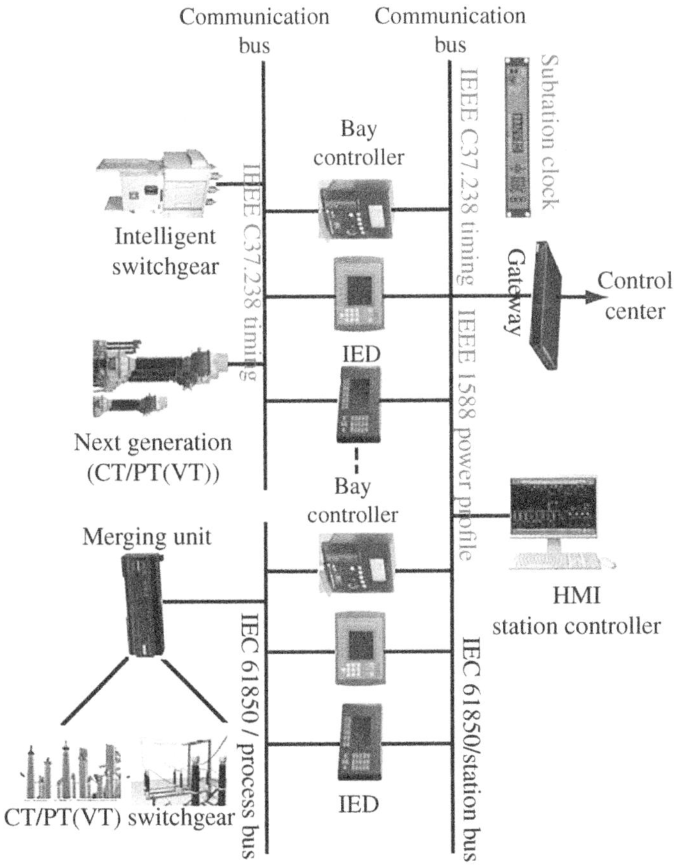

Figure 10.1 An example of using IEC protocol.

or monitoring within substations [14]. The interoperability is the aptitude of two or more IEDs from the same manufacturer or different manufacturers to exchange information and use it for the proper execution of the specified functions [15, 16]. This standard gained popularity and continues to spread around the globe [17]. Based on the flexibility of the standard, IEC 61850 can be used to cover many applications and standardized the different parts of the IEC 61850 series.

10.3 IEC 61850 Protocols and Substandards

10.3.1 IEC 61850 Standards and Classifications

IEC 61850 consists of different parts that define different aspects of IEC 61850, as shown in Table 10.1. These parts constitute from general information such as the introduction and overview in part 1 to conformance testing in part 10. It focused on

Table 10.1 IEC 61850 substandard.

Part	Designation	Definition and specification
1	Introduction and summary	"Presentation to IEC 61850 and a general description of all the parts"
2	Vocabulary	"Specific terms of SAS"
3	General conditions	"System operations based on quality requirements"
4	Project management and system	"Service engineer requirements"
5	Requirements of communication for characteristics and models of devices	"Aspects of virtualization and performance specifications"
6	Description in the language of the configuration for the communication in electrical substations related to IEDs	"File format to describe the system configuration and the relationship between the devices"
7	A fundamental communication system for feeder equipment and substation	"Communication requirements and specifications"
7-1	Models and principles	"Principles of the information and communication mapping scheme"
7-2	Abstract communication service interface (ACSI)	"Cooperation of various devices"
7-3	Logical nodes (LN)	"Common data classes and common attribute type used in substation application"
7-4	Appropriate logical node classes and data classification	"Classes of data on syntax and semantics"
8	Specific communication service mapping (SCSM)	
8-1	Mappings to MMS (IEC 9506-1and IEC 9506-2) and IEC 8802-3	"Communication mapping for the entire system"

Table 10.1 (Continued)

Part	Designation	Definition and specification
8-2	Mapping to extensible messaging and presence protocol (XMPP)	Mapping to MMS, XML, and XMPP
9	Specific communication service mapping (SCSM)	
9-1	SMVs through a serial unidirectional multi-drop peer-to-peer link	"Peer-to-peer unidirectional communication services mapping"
9-2	SMVs through IEC 8802-part 3	"Specific communication service mapping for a type of bus"
10	Compliance test	"Implementation of compliance testing techniques and techniques to measure the declared performance parameters"
	Communication networks and systems for power utility automation	
7-420:2009	Part 7-420	Basic communication structure Distributed energy resources (DERs) LNs
90-1:2010	Part 90-1	Use of IEC 61850 for the communication between substations
7-410:2012+ AMD1:2015 CSV Consolidated version	Part 7-410	Basic communication structure Hydroelectric power plants Communication for monitoring and control
7-510:2012	Part 7-510	Basic communication structure Hydroelectric power plants Modeling concepts and guidelines
90-5:2012	Part 90-5	Use of IEC 61850 to transmit synchrophasor information according to IEEE C37.118
90-4:2013	Part 90-4	Network engineering guidelines
90-7:2013	Part 90-7	Object models for power converters in DERs systems

Table 10.1 (Continued)

Part	Designation	Definition and specification
90-12:2015	Part 90-12	Wide area network engineering guidelines
80-1:2016	Part 80-1	Guideline to exchanging information from a CDC-based data model using IEC 60870-5-101 or IEC 60870-5-104
80-4:2016	Part 80-4	Translation from the COSEM object model (IEC 62056) to the IEC 61850 data model
90-2:2016	Part 90-2	Using IEC 61850 for communication between substations and control centers
90-3:2016	Part 90-3	Using IEC 61850 for condition monitoring diagnosis and analysis
90-8:2016	Part 90-8	Object model for E-mobility
7-500:2017	Part 7-500	Basic information and communication structure; the use of LNs for modeling application functions and related concepts and guidelines for substations
90-10:2017	Part 90-10	Models for scheduling
90-17:2017	Part 90-17	Using IEC 61850 to transmit power quality data
8-2:2018	Part 8-2	Specific communication service mapping (SCSM); mapping to extensible messaging and presence protocol (XMPP)
90-6:2018	Part 90-6	Use of IEC 61850 for distribution automation systems

three main concepts: functional modeling of the application automation domain (Part 5), substation automation system (SAS) data modeling (7), and protocols and communication services such as system configuration language (SCL) or configuration description language based on XML (6) [18, 19]. The most important contents are found in Parts 7-x. Recently more new parts are published. Ref. [20] explains each part by its scope and purpose.

10.3.2 Basics of IEC 61850 Architecture Model

IEC 61850 defines a series of architectural objects designed to structure the intelligence of protection, control, monitoring, and automation functions. The main architectural construction adopted by IEC 61850 is the abstracting of these data objects (data, data attributes) and service models, which are found in IEC 61850-7-x, using the Abstract Communication Service Interface (ACSI) in part 7-2 [21]. The abstract models can be mapped to arbitrary communications protocols [22]; thus, the system will be compatible with future developments in communication technology [23].

10.3.3 IEC 61850 Class Model

The IEC 61850 models provide two approaches based on the specifications of [24, 25]:

- A basic model for defining the specific information models of the substations contained in IEC 61850-7-3 and IEC 61850-7-4.
- Information exchange service models.

IEC 61850 defines each function within the equipment of the substation (transformer, circuit breaker, protection function, etc.) by a logical node (LN). Each LN composed of data sets, data attributes, and associated communication services. Unified Modeling Language (UML) model of IEC 61850 comprises the following number of LN classes, data objects (attributes), enumerations, and abbreviations as shown in Table 10.2 [26].

There are ninety-one a total of 468 LN classes divided into thirteen logical groups [27, 28]. The 1st letter of the LN identifies the group, as illustrated in Table 10.3.

Each LN class consists of seven categories of data types, such as status information, measured information, etc. Each data class is composed of 3662 different data attributes that have a specific name, a specific type, and several purposes [29].

Figure 10.2 illustrates the hierarchical model of the data structure of IEC 61850. It consists of five layers from the IED or "physical device" to the data attribute.

Various LN classes can form the virtual analog of a physical, electronic device called a logical device (LD). As an illustration, "MMXU: is the power, currents,

Table 10.2 IEC61850 UML model.

	WG10	WG17	WG18	JWG25	Other	Total
LN classes	279	118	51	15	5	468
Attributes	1847	966	646	152	51	3662
Enumerations	133	81	19	39	1	273
Abbreviations	976	81			8	1065

Table 10.3 Indemnification of "logical nodes" (LNs).

Logical group	Identification	Number
A	"Automatic control"	8
C	"Control (supervisory)"	5
G	"Generics"	3
I	"Archiving/interfacing"	4
L	"Logical node systems"	3
M	"Measurement/metering"	8
P	"Protection system (safety and security)"	28
R	"Protection related (common SCADA)"	10
S	"Sensors (detectors/actuators)"	4
T	"Transformers for instruments"	2
X	"Switchgear(circuit breaker/switch)"	2
Y	"Transformers"	4
Z	"Different equipment"	15

voltages, and impedances in a three phase system," and "XCBR: is a circuit breaker" (see Figure 10.3).

The number of LN increases with the acceptance of the standard [30, 31]. SCL describes and enables the configuration of IEDs in terms of functionality communication addresses and services (reporting). So, the configuration of the system devices can be done before use and additionally in the case of including new devices to the system [32].

Indeed the SCL is an important part of IEC 61850. It allows the formal description of the relationships between the utility automation system and the substation/switch yard (process). Its principal objective is to allow the

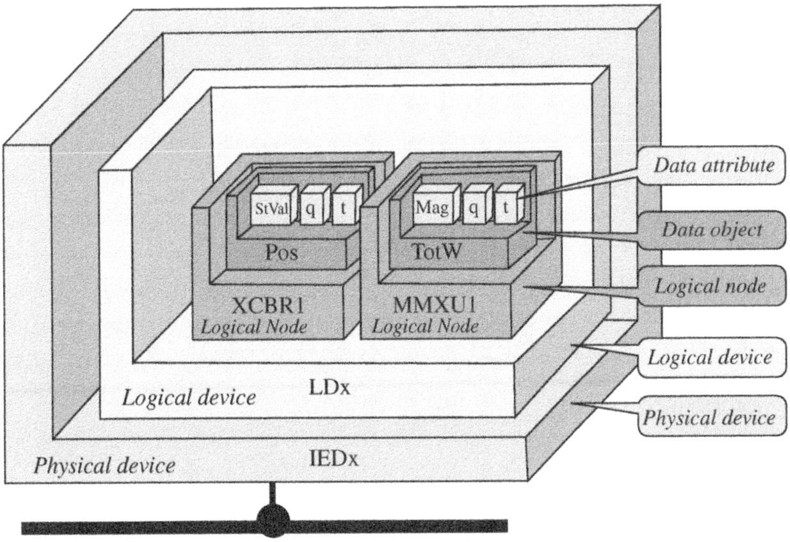

Figure 10.2 Data structure of IEC 61850. Source: IEC [34] / with permission of International Electrotechnical Committee.

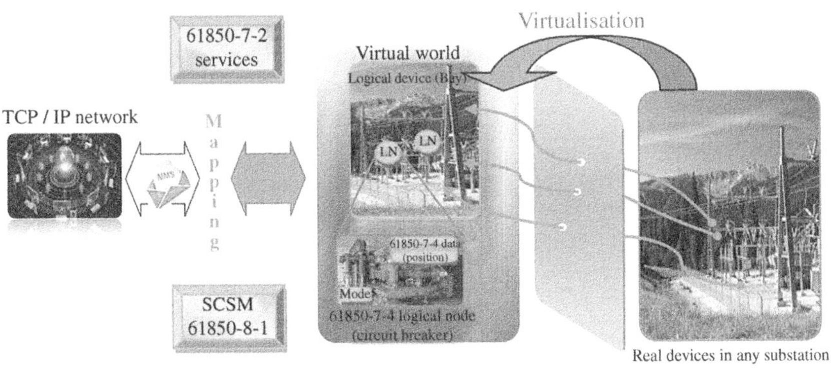

Figure 10.3 Networks and communication systems in substations.

interoperable exchange of communication system configuration data between the configuration tools of both IED and system from different manufacturers. This configuration language is based on the XML version 1.0 [32].

This example is inspired from one of the editors Klaus-Peter Brand who was member of ABB company:

```
<Substation Ref="">
<VoltageLevel Ref="E1">
<Bay Ref="Q1">
```

```
<Bfunction Ref="">
<Device Ref="QA1" Type="CBR">
<Connection NodeRef="L1"/>
<LNode Ref="1" LNClass="CSWI"/>
</Device>
<Device Ref="QB1" Type="DIS">
<Connection NodeRef="L1"/>
<LNode Ref="2" LNClass="CSWI"/>
</Device>
</Bfunction>
</Bay>
</VoltageLevel>
</Substation>
```

Here, "Described" is a substation with the bay E1Q1, the circuit breaker QA1, and the isolator QB1, both electrically connected in connection node L1. The controller is represented by LN CSWI controlling both switches [33].

10.3.4 IEC 61850 Logical Interfaces (Functional Hierarchy of IEC 61850)

The IEC 61850 consists of a model regarding a substation communication interface system, as seen in Figure 10.4. It divides substation communications into three levels hierarchically station, bay, and process levels.

- **Station level**: As seen in Figure 10.4, interfaces 4 and 5 depict the communication between the bay and the process levels, and interfaces 1 and 6 represent

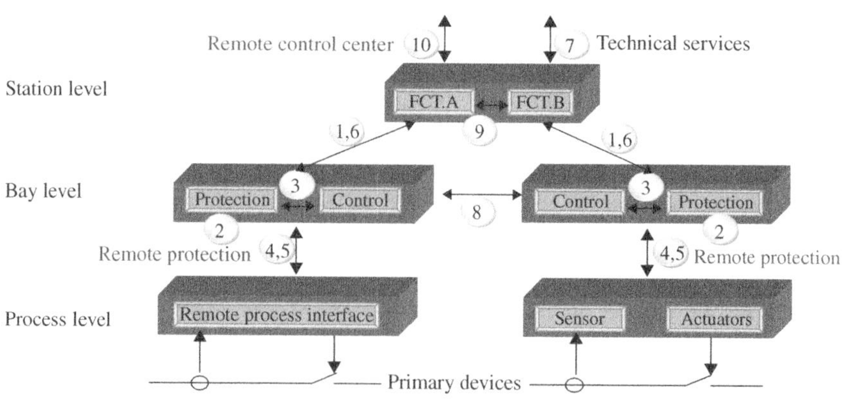

Figure 10.4 Functional hierarchy of IEC 61850. Source: IEC [34] / with permission of International Electrotechnical Committee.

the communication among the bay and the station levels. The latter includes human–machine interface and engineering workstations additionally gateways to link substation control center to "wide area network" (WAN) [22]. It communicates with the bay level within interfaces 1 and 6 to exchange protection data and control data, respectively. Besides, interface 9 depicts the sharing of data within the station level [35].

- **Bay level**: The bay level (local control rooms), as defined by IEC 61850 requirements, is an intermediate control location between switchgear boards and the main control cabin of the substation [36, 37]. It includes protection and control IEDs of various bays, e.g. transformers, circuit breakers, capacitor banks, etc. Several functions within a single bay unit can share data, as shown by interface 3. The voltage signals and currents measured by PT/CT are digitized by merging units (MUs) and will be sent to the bay level via the Ethernet network from the process level [36].
- **Process level**: The process level has been connected to the bay level through interfaces 4 and 5. These logical interfaces offer the possibility of exchanging control commands and information data [38]. The process level consists of switchyard equipment, which includes transformers, circuit breakers, switchgears, potential transformers, current transformers, actuators, and MUs called primary equipment. The primary objective of this level is to collect data from electrical processes and perform switching operations [38]. The communications of these three levels are performed over the station and process buses. In SAS, this network operates under the IEC 61850 platform using different communication services for the communication within the bay level and station level devices [37].

10.4 IEC 61850 Features

IEC 61850 communication service supports several services including Generic Object-Oriented Substation Events (GOOSE), ACSI mapped to Manufacturing Message Specification (MMS), Sampled Values (SVs), and web services. These different services have different real-time requirements. They can operate on TCP/IP networks or substation LANs using high-speed switched Ethernet to achieve the required response times of less than 04 ms for protection relaying [39]. Figure 10.5 shows some of these communication services supplied by IEC 61850. SV and GOOSE communications are directly mapped to the Ethernet layer subsequently; they dispose processing of any intermediate layers and increase the performance (time-critical services) while MMS is mapped based to the standard TCP/IP due to its time unlimited [41].

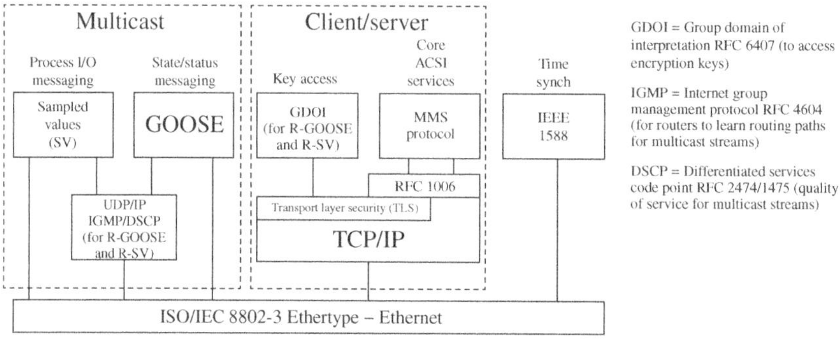

Figure 10.5 IEC 61850 communication services. Source: Based on Personnel communication [40].

10.4.1 MMS

MMS is an international standardized messaging system (ISO 9506) supporting the transfer of real-time process data indications, control operations, and report notification between networked devices and computer applications (vertical communication). The MMS defines the following [42]:

- A set of standard objects that must exist in each device, on which operations such as reading, writing, event reports, etc. can be performed. The Virtual Manufacturing Device (VMD) is the fundamental object, and all other objects such as journals, domains, variables, files, etc. are included in VMD.
- A series of standard messages exchanged among a client and a server stations to monitor and control these objects.
- A set of encryption test for mapping these messages to bytes and bits while transmitting.

10.4.2 GOOSE

GOOSE is the acronym for Generic (any records) Object-Oriented (information from IED 61850 information model) Substation (complete substation) Event (function or occurrence based sending) and is a multicast service used for protection and automation of electric structures in the substation. In this utility, GOOSE messages are passed horizontally peer to peer between IED 61850 IEDs [43].

An IED (publisher) creates a message that is destined to a group of IEDs (subscribers) simultaneously [44].

At the publisher part, the values are written in the local buffer while subscribers read the values of the local buffer at the receiving part. The communication system deals with the update of the local buffers of subscribers. Indeed the relevant objectives for a GOOSE message are strict, and it is not allowed to spend more than 4ms from the moment an event occurs when the message is received [45].

The IEC 61850 has been developed to apply purposes that require intersubstation communication as tele-protection, which needs the use of GOOSE message through WAN for instant broadcasting of control signals [46, 47].

10.4.3 Sampled Measured Value (SMV) or SV

SV messages are exchanged in a peer-to-peer publisher/subscriber mechanism (unicast or multicast models), and SV is used to transmit digitized analog data such as temperature, current transformer, and voltage transformer that come from instrument transformers. The publisher mechanism (MU) registers the obtained value in an internal buffer, and the subscriber mechanism (IED) reads the value and transfers it to its buffer. A timestamp is added to each obtained value to allow the subscriber mechanism to follow the correct value sequence [37, 48].

A single objective can request an individual sampling rate that can be freely selected to meet their requirements, for example, 80 samples per nominal period for a protection application whose sampling rate is 4 kHz/s for a frequency of 50 Hz. It allows 01 set of samples to be sent in a single SV packet every 250 μs and 256 samples per nominal period for a measurement application in which it is possible to send 08 sets of samples in a single SV packet every 78.125 μs [49]. The IEC 61850-5 divides messages into seven classes, from class 1 covering speedy messages to class 6, shown in Table 10.4.

GOOSE messages are class 1, or 1A speedy messages must have back-to-back transmission time of 3 ms appropriate for substation protection and crucial control, while the required precision of time synchronization dictates the performance requirement of the type 6 message. The IEC 61850 communication standard requires that the precision of the time synchronization be 10 times faster than the required timestamp precision. The message speed must be fast and accurate enough to synchronize the IED clocks with an accuracy of 0.1 ms so that the timestamps of the IED data can be accurate to 1ms [50, 51].

Since GOOSE and SV messages might be transmitted over WAN, cybersecurity (IT security) requirements are considered in the technical report IEC TR 61850-90-5: power systems management and associated information exchange. These requirements defined the need to extend GOOSE and SV as Routable GOOSE (R-GOOSE) and Routable-SV (R-SV) by adding network and transport layers.

In current years, some review studies has been published in the domain of application of SV and GOOSE message services. Lee et al. have developed IEC 61850 for PMU interface using MMS and GOOSE message services to upgrade the performance of monitoring system [52]. Xu et al. have employed synchrophasor

Table 10.4 Type of message with distinct communication times.

Type of message	Identification	Type of communication services	Transmission time (ms)	Application
1, 1A	Time-critical messages	GOOSE	3	"Protection that contains command, data, or message, for example, close, trip, etc."
2, 3, 5	Used to support core IEC 61850 services	MMS	100–1000	"Refer to the messages between substations for automatic functions and to operate messages, not time critical, referring to the type of slow response of a person"
4	Raw message	SV	3	"Samples: raw data from digitizing transducers and digital instrument transformers independent from the transducers technology (optic, magnetic, etc.)"
6	The message used for time synchronization	Simple Network Time Protocol (SNTP)	Accuracy	

data mapping to SV message service [53]. However, not any one of these studies addressed R-GOOSE and R-SV presented in part 90-5.

Indeed, in the analysis of the 2003 blackout in the Northeastern United States, it was indicated that a need for a well-known synchrophasor measurements is fundamental to prevent such great trouble in the grid [54].

10.4.4 R-GOOSE and R-SV

Routable or routed, as indicated by its name, is receiving messages and transmitting them to the expected address. As mentioned before, the R-GOOSE and R-SV messages are routable over User Datagram Protocol/Internet Protocol (UDP/IP). There are used for protection systems in distribution automation (DA) and also for the integration of distributed generation (DG) and distributed energy resources (DERs) [54].

The concept of synchrophasor data transmission in the cases of IEC 61850 has been reviewed and studied in Refs [55–59].

Apostolov has described in detail the concept of R-GOOSE and its application in DA. He showed the distinction between GOOSE and R-GOOSE messages and their features and data models. On other hand, he has not introduced appropriate communication middleware and has not addressed network issues [55].

Youssef et al. have proposed a data-centric communication framework for multicast IEC 61850 R-GOOSE message over the WAN in modern power system. The effectiveness of this approach has been evaluated by experimental results obtained with an actual hardware-based test setup developed in the Florida University smart grid testbed [56].

Firouzi et al. proposed a method for using IEC 61850 R-SV and R-GOOSE messages to transmit synchronized phasor measurements over WANs to enable wide area monitoring, protection, and control communications. However, the cybersecurity features were not treated [57].

Madani et al. have implemented a hybrid protocol synchrophasor system using R-SV message service for data transmission and C37.118 for carrying the configuration of the PMUs. However, no security mechanism has been mentioned in this research [58].

Ustun et al. have implemented secure R-GOOSE and SV messages based on IEC 61850-90-5 to mitigate security attacks. They have developed a novel toolbox called R-GoSV to construct phasor measurement unit (PMU) messages with cybersecurity mechanisms. Thanks to this, custom messages have been communicated in the network to examine their effectiveness [59].

In the rest of this chapter, the application of synchrophasor systems will be treated briefly since this subject has been already done and presented in details in Refs. [60–62].

10.4.4.1 Application in Transmission Systems

The electric power collection data in real time becomes a requirement in the creation of smart grid since smart grid data needs complex treatments according to its nature and distribution. Thanks to the PMUs with their synchronized real-time measurements through the grid, this allows increasing power quality and continuity of service and system reliability.

Review of particular applications in transmission given here consists of wide area control and monitoring, state estimation (SE), fault location, and protection design scheme, and other applications in distribution such as load modeling, harmonic estimation, and islanding detection are also presented briefly.

Wide Area Control and Monitoring To improve the power quality and secure the electric grid, it is vital to incorporate Wide Area Monitoring and Control (WAMC) system. This control can act as either prevention or correction. The WAMC requires modern sensors such as voltage transformer and current transformer for voltage and current measurements to provide situational awareness and information and communication technology (ICT) systems such as global positioning system to provide synchronized time signal or fiber-optics communication network to communicate data. Some of the examples of WAMC are frequency, voltage stability, and oscillation. Significant works have been dedicated to the subject of progress of WAMS applications to power systems.

Zima et al. have developed algorithms and their design for applications of the WAMC system, i.e. frequency instability evaluation, vibration detection evaluation, voltage stability evaluation, and line temperature monitoring. They showed that such a system could be used to enhance the security of power systems by providing real-time monitoring of key variables and enabling emergency control [60].

Zhu et al. presented a case study showing the dependency of WAMC application reliability on the supporting ICT system architectures, focusing on data delays and losses. The authors proposed a generic ICT architecture that enables large-scale implementation of WAMC systems and pointed out that supporting ICT system with inappropriate design risk devaluing WAMC systems [61].

Korba and Uhlen devoted themselves to two different approaches, the first based on an adaptive Kalman filter and the second based on a multivariable nonadaptive SI, for detecting critical oscillation modes. They used rapidly sampled signals acquired with phasor meters and a large-scale monitoring platform. The goal was to provide the operator with information about the occurrence of oscillations in the monitored parts of the network in (near) real time. They found that the (near) real-time monitoring system was able to quickly and accurately identify the amplitude and frequency of the dominant oscillation mode. Furthermore, identifying the modal damping was more difficult in real-time applications, especially when the magnitude of the oscillations was in a range comparable with the random load fluctuations [62].

State Estimation SE is a method to estimate the state of power system that is represented by scalar data, viz. voltage magnitude, voltage angle, current quantity, and active and reactive powers. These raw data are corrected by SE to reduce their errors, to compensate for nonexistent values, and to eliminate inaccuracies data. The SE is used to carry out the normal operation of power system and is also used as emergency control action to rescue the power system from blackout and to restore the system following an eventuality [63].

Gomez-Exposito et al. proposed a new paradigm based on a multilevel framework that facilitates the seamless integration of existing SEs designed to function at different levels of the modeling hierarchy to achieve very large-scale monitoring of interconnected power systems [64].

Chakhchoukh et al. introduced a novel approach that estimated the power system state in the presence of some PMU measurements (hybrid static SE). This approach allowed to improve the SE and to provide more accurate confidence intervals by considering important information, namely, the space and time correlation available in the recorded PMUs.

Göl and Abur have presented a thorough and detailed performance comparison between least absolute value (LAV) and weighted least squares (WLS) estimators. They showed that the LAV estimator is a statistically robust and computationally competitive estimator for those energy systems that are fully measured by PMUs [65].

Fault Location and Protection Design Scheme Power system is regularly exposed to faults created by different causes i.e. faulty materials, disasters, severe weather, etc. In the case of a faulty power system, either the customers are deprived of electrical power and the system is at contingency, or the whole demands of the system are till being supplied and the system security is critically jeopardized .Therefore, accuracy and fast detection and network fault location were historically recognized as a vital task in the system maintenance, repair, and inspection actions. Thanks to the PMU, the data such as voltage magnitude, voltage angle, etc. are transmitted to the nodes at the transmission side. PMUs can be applied either in fault location or in protection design scheme.

O'Brien et al. prepared a report that provided general background information on synchrophasor measurement technology and several important aspects of the technology's application, such as communication requirements, performance and interoperability requirements, time source considerations, and other related topics. The report also described protective relay applications using the PMU, such as power fluctuation detection, load shedding, line reclosing selectivity, power system analysis, synchrophasor-assisted black start, islanding protection for DG, automatic generator shedding, communication channel analysis, voltage and current phasing verification, and distance to fault [66].

Khorashadi-Zadeh and Li have proposed an alternative scheme for adaptive auto-reclosing with a secure and variable dead-time control algorithm in the presence of transient faults via secondary arc extinction detection. The proposed algorithm calculated the post-fault phase voltage at the measurement point after the opening of the circuit breaker and used the post-fault phase voltage to classify the transient fault and the permanent fault and to estimate the arc extinction time [67].

Nayak et al. proposed a synchronized phasor measurement-based wide area backup protection scheme that uses angle information between voltage and current at both ends of a line for faulted line identification (FLI). They showed that the compensated and uncompensated line schemes were accurate and also had the ability to distinguish the fault from load disturbances and power fluctuations and that method could overcome the limitations of the conventional backup protection scheme [68].

Brahma proposed an iterative algorithm using the properties of the impedance matrix of the bus with positive sequence Zbus. The algorithm first identified the faulted section and then localized the fault on that section. This algorithm was applied to the data obtained from electromagnetic transient program simulation of a multiterminal transmission line. The simulation results showed that this method was characterized by the fact that it worked with very high accuracy for all types of faults at different fault locations and was practically immune to fault resistances [69].

Izykowski et al. have presented a new method for locating faults on three-terminal power lines. They have showed that the set of fault location input signals used allows accurate determination of the total fault courant flowing through a fault path resistance [70].

10.4.4.2 Application in Distribution Systems

Load Modeling Load is among the essential component in the power system modeling. Load models can be classified as static and dynamic types from the modeling prospect [10]. Static model is only based on the current state of the system, while dynamic one is a function of present and precedent states. To ensure pragmatic simulation results for tremendous variety of power systems, it is necessary to have an accurate load models.

The load modeling difficulties is primarily due to the unpredictability resulting from lacking data on load component's detail and climate seasonal fluctuations. For this reason, the increasing availability of fast response PMU in power systems has incited more research on their application in load modeling studies [71, 72].

Kosterev has conducted an approach for generator model validation based on the PMUs taken at the point of interconnection. These PMUs measured bus frequency, voltage phasors, and current phasors in Dalles, John Day, and Grand Coulee powerhouses. The PMU data was employed for synchronous generator dynamic model validation [71].

Kennedy et al. have outlined the use of PMU records to validate models of wind farms based on fixed-speed induction generators (FSIGs) during frequency transients. They have showed that model accuracy varies as a function of the size and specific type of wind turbine generator and also as a function of the level of wind generation output [72].

Harmonic Estimation Harmonics, which are generally generated by nonlinear loads (drive, arc furnace, rectifier, starting motor, etc.) linked to the power system, represent polluting sources on power quality diagnostic. These harmonics can induce damages in the power system such as destruction of circuit breaker, mechanical stress in asynchronous machine, or accuracy loss of measurement device and so on. Harmonic Measurement System (HMS) was performed and used by United States since 1992 using PMUs.

In fact Carta et al. have presented the design and characterization of a PMU based on a PXI modular system for measuring harmonic synchrophasors in different nodes of a distribution network. They have implemented suitable measurement procedures using high-accuracy GPS receivers and general acquisition hardware. They showed that the PMU prototype was capable of evaluating the harmonic synchrophasors with an accuracy that meets the requirements of most distribution network applications [73].

Melo et al. have presented a methodology for HSE for unbalanced three-phase distribution network using PMU, taking into count the presence of harmonic sources connected to the system. They found that the harmonic currents/voltages and THD profiles of current and voltage were satisfactorily estimated [74].

Islanding Detection Islanding detection is among the relevant features of interconnecting DGs to the power system. Islanding detection modes can be generally divided into two modes: remote mode and local mode. An islanding is the circumstance in which DGs remain to supply a location even over electrical power grid is absent or not available. This might be intentional or unintentional. Islanding can deteriorate the equipment, influence the reliability of power system, and cause danger to the maintenance operator's life. The IEEE Standard 1547-2008 defined a required time of two seconds for detection of the unintentional islanding [75].

PMU can be useful to detect islanding event thanks to the fast secure communication; the disconnection of the DGs can be carried out in two seconds from the network.

Borghetti et al. described the performances of a PMU model based on a synchrophasor estimation algorithm designed for monitoring active distribution networks and its experimental application during some intentional islanding and reconnection tests of an urban medium-voltage network. They have developed a PMU that requires very low values of total vector error and includes particularly low values of the phase errors of the synchrophasor estimates. Three of these PMUs were then used to monitor experimental tests conducted to evaluate the ability of an urban distribution system to operate autonomously when fed by a local 80 MW combined cycle power plant. The information given by the installed PMUs greatly facilitated the operator's maneuvers and proved useful in developing an improved control and management system of the active distribution system [76].

Best et al. have investigated the detection of an islanded network created by unintentional islanding. They have used PMUs for synchronization of islanded network. The voltage phase angle was used to identify DG in the same island by comparing PMUs at the supervisory controller [77].

10.4.5 Web Services

A web is a service dispatched out via an IED to every other IED communicating with each other through the World Wide Web. In a web provider, the HTTP is used for machine-to-machine communication, in particular, to transfer files in a machine-readable format, such as XML. Indeed the web provider is not unusual in Internet and cloud programs, and it receives more consideration for the realization of smart grid systems [78]. The part IEC 61850-8-2 specifies a recent typical mapping of the data models and services to web services. The process related requirements for web-based and established access to the IEC 61850 data models.

10.5 Relevant Application

IEC 61850 has gained a wide range of application possibilities exceeding by far present applications. This is available for the immediate future in terms of integration, automation, and control of electrical substations. Destined for use at distribution, transmission applications in substations and on customer part of the meter level, consequently, it extends to new topics such as advanced metering infrastructure (AMI), distribution management system (DMS), energy management system (EMS), electric vehicle (EV), and several types of DER.

IEC 61850 has been widespread beyond the field of SAS to cover DERs, EVs, and the communication to control center. As a result, it can substantially be used to support metering services within the distribution network, starting with the support for AMI.

10.5.1 Substation Automation System (SAS)

The substation is a key component of the smart grid, which facilitates the transit of energy electric [79, 80]. It plays a fundamental role in the control and monitoring of energy flow and provides interconnection between power plants, transmission

Table 10.5 Basic functions of an SAS.

Fundamental functions	Operations
Monitoring	- Status monitoring of switchgear, control and protection equipment, transformer and load selector, etc. - Monitoring and supervising of measurement input data (reactive and real power, current, voltage, frequency, etc.) - Facility failure/event and system disturbance record
Control	- Transformer tap raise/lower commands and control close/open commands to switchgear - Interlocking and synchronism check - Load reduction, reactive and active power, voltage regulation, etc.
Protection	- Transformer and overhead line and cable, busbar, etc. - Overcurrent, distance protection, earth fault relays, etc.
Configuration	- Configure settings

and distribution grids, and customers. These fundamental functions of an SAS can be classified as listed in Table 10.5.

Besides, SAS affords supplementary information and capabilities that can be used to further save the operation of substations [23, 81–83]. The characteristics of IEC 61850 are seemingly small but can have a tremendous impact on SAS [1, 84, 85] such as the following:

- Usage of a virtualized LDs, LNs, and Common data class models.
- Interoperability between devices of several manufacturers within the power system and possibility to configure the system with the available configuration system tools, regardless of the manufacturer's support on the site.
- All object names normalized and specified in an energy system background.
- Flexible and extensible functions and reconfiguration by the self-description in a normalized way.
- A secure and reliable general system that uses several techniques allowing for flexible data transfers.
- Reducing the overall system costs for operation and maintenance.
- An open scheme for the automation, protection, and data exchange from the usage of the international representation for all the objects of the energy system and common technology infrastructure to dispose of supply ambiguities.

10.5.2 Energy Management System (EMS)

The demand for electricity is growing rapidly. Therefore, the availability of electricity is a critical issue. Smart grid permits the control of energy and real-time stability between the supply of electricity and the demand side [86]. EMS is defined in Ref. [87] as a control tool that controls the energy flows between the main network, the DER, and the loads to provide stable, reliable, and sustainable operation of the micro-grid and other operational objectives, such as reducing cost and fuel consumption [88–90].

It is also responsible for re-synchronizing the system all through transition among grid and island modes. For EMS, it is possible to identify two main approaches such as decentralized and centralized control, with several hierarchical controls [91–94]. In other words, EMS is a system referred to SCADA used by grid operators to manage and optimize the performance of power network applications and to the generation control and scheduling applications (energy efficiency). It can supply sub-metering, metering, and monitoring functions that permit gathering data and making more informed decisions about energy activities [95, 96]. SCADA is one of the fundamental missions of the SAS, which includes the following [97, 98]:

- Remote and local operation of the switchgear and different high-voltage system.
- Acquisition of switchgear data and measurands of power systems.
- Managing of event and alarm.

In a substation, the client is generally the station controller, the server is an IED as a protection relay, and the C/S communication is used primarily for the classic exchange of SCADA data. Two functions for condition monitoring and SCADA are specified as functions for supervising [99]:

(1) **Status value**: any modification in a status value can issue a message (report) in a SCADA system or another system or can publish the modified value in a record (the record be read later).
(2) **Analog value**: absolute boundaries could be supervised for violations (low, high, low–low, high–high, etc.) or relative boundaries (deadband and % difference in value). The boundary violations can be reported or recorded.

The SCADA utility is associated with the human operation of the network and carried out employing remote or local operators. The operators use a client system as a terminal to manage the electrical equipment through the control units that are connected to the electrical equipment and act as an SCADA system server.

The control unit for connection to the network exchanges state data, alarm for the transformer, circuit breaker, protection device, and the measurement data, including real power, reactive energy, active power, etc. [100]. Lately, automated fault location, isolation, and supply restoration (FLISR) uses the system of SCADA as "Eyes and Arms" to collect data about failures and carry out the necessary control actions (closing and opening of switches). In the most generally known structure, FLISR is a subsystem of a centralized DMS [101].

10.5.3 Distribution Management System (DMS)

A DMS is a set of programs that may supervise and manage the whole distribution network efficiently and reliably. It serves as a decision support system to help the field operating personnel and control room with the supervising and control of the electric distribution system. Enhancing the reliability and quality of service concerning minimizing interruptions, reducing downtime, and maintaining acceptable levels of frequency and voltage are the principal products of a DMS.

The development in automated metering initiatives, DA, and substation automation in the industry has created additional progress to establish the DMS as a solid foundation to exploit these aspects of the new intelligent network.

Mainly, the four peripheral equipment covered by DMS are dispatching and control center (like OMS, SCADA, and GIS), field devices (such as relays, remotely controlled switches, and other IEDs), integration with enterprise applications (like customer information systems, enterprise resource planning (ERP) platforms, and human resource systems), and finally communication systems (as WAN and LAN technology, radio, fiber optics and RTUs) [102].

The development of DMS by enhancing IEC61850-based feeder IEDs of distribution switchgear was carried out in [103]. The authors have proposed nine types of services, defined in the IEC 61850 communication services, to be used in the feeder IEDs. GOOSE service is used to exchange event messages for protection coordination. Ali and Hussain [104] demonstrates the modeling of information and services needed for control, protection, and management of distribution systems with integrated DERs.

They have used IP tunnels and mapping over IP layer for transferring IEC 61850 messages, such as SVs and GOOSE over distribution system WAN.

Senke et al. [105] evaluates the applicability of IEC 61850 from the point of view of LN and DO for communication in DA and building EMSs. They found that most data encountered in such communication can be represented as standard LNs and DOs, but multiple data elements cannot be represented simply as the same. DMS providently optimizes network resources, minimizes peak demand, and helps distribution networks to supply electricity more reliably, efficiently, safely, and economically [106]:

10.5.3.1 Feeder Balancing and Loss Minimization Distribution

Switching optimization automatically determines the optimal system topology to balance the load on the feeders and, in combination with loss minimization, eliminates abnormal operating conditions. It determines the optimal state of the existing switchgear and proposes the positions for the new open feeders in the system.

10.5.3.2 Voltage/VAR Optimization (VVO) and Conservation Voltage Reduction

Voltage/VAR Optimization (VVO), as a part of DMS, is among the fundamental applications used for managing in distribution network operations. As such, it lets in the management of voltages and reactive energy flows in the distribution network. It works centrally throughout the considered subnetwork, instead of locally, as is possible with local controllers and assists distribution operators to reduce system losses, peak demand, or energy consumption using conservation voltage reduction (CVR) techniques.

10.5.3.3 Fault Location, Isolation, and Service Restoration

FLISR locates the section of the network that is isolated due to a forced outage and provides the operator or dispatcher with information about the affected customers. Furthermore, it used to restore the supply to all customers without power.

10.5.4 Distribution Automation (DA)

It is a vital element of smart grid and is primordial to increase performance and availability of the grid, reduce the duration and frequency of interruptions, enhance recovery time, and reduce operating costs. IEC 61850-90-6 [6] is the basic standard in the domain of smart grid for DA.

Indeed interoperability of all advanced sensors, SCADA systems, and electronic controllers that are part of the DA system require IEC 61850 communication to cover all the components including the end consumers and also to solve plug-and-play problems. In this context, Mohaghegh et al. [107] discusses the application of IEC 61850 in DA and provides some concrete examples.

In [108], an open communication architecture for DA based on IEC 61850 is suggested and discussed how to realize the specific features of DA. The configuration solution for IEC 61850 implementations in distribution automation systems (DASs), especially for the distributed applications, is suggested in [109].

The authors have used the new SCL models to describe the distribution network technology and have presented a new "configured IED description" (CID) file that includes the topology, communication system, and functional metering devices and data models as required for the distributed application. Yip et al. [110] presents an application of IEC 61850 for distribution network automation with decentralized control, which describes in detail the mapping to LNs and the creation of new LNs and various configuration scenarios.

A muti-agent smart grid automation architecture based on IEC 61850-61499 intelligent LNs was presented in Ref. [111]. It was shown that this approach can achieve a self-healing network through collaborative fault location and power restoration.

In another study [112], the authors evaluated communication technologies for IEC 61850 based DAS with DERs. They have simulated the communication networks for a typical DAS with DERs using OPNET software. They showed the performance of wired and wireless communication systems for different messages, such as GOOSE and measured values between DERs and DAS.

Tobar et al. [113] presents the analysis of the basic characteristics and topologies for DA suitable for using renewable energy as a power source. DA is considered as a real-time automated remote supervising of main distribution application system along with capacitor banks, auto-reclosers, distribution transformers, ring main units, fault passage indicators, and sectionalizers to integrate it with SCADA/DMS equipment [114].

A few common DA utilities have been introduced depending on integration with IEC 61850. These consist of service restoration (SR), VVC, and finally fault detection and isolation (FDI) [107, 115]:

10.5.4.1 Voltage/VAR Control

In general, the goal of the VVC problem is to limit the energy losses in the grid, flatten voltage profile, and minimize the peak demand. This goal is obtained by adjusting the tap positions of the voltage regulating transformers and the size/state of the shunt capacitors along with the feeders, including that of the on-load tap changer at the substation.

The optimization problem is subject to operational constraints of the two main control elements of the VVC problem, namely, voltage regulator tap position control and shunt capacitor.

10.5.4.2 Fault Detection and Isolation

FDI aims to detect and discover a fault in the grid, resulting in isolating it from the rest of the grid in a very short time. Therefore, the FDIs contribute to the complete safety of the system by minimizing the impact of a power failure on the grid. It must ensure that the fault is isolated with a minimum amount of unserved load due to disturbance. This mechanism depends on a large extent on the protection scheme used in the power supplies. In general, two main protection systems are currently in use throughout the world, namely, the distributed protection scheme and the centralized protection scheme.

10.5.4.3 Service Restoration Use Case

After the occurrence of a failure in the distribution network and the subsequent separation from the rest of the network, not only the customers connected to the faulty circuit are initially affected by this interruption, but also other customers, e.g. those at the downstream fault location, remain without power. SR algorithms are then applied in the DMS to provide power to residential customers while the fault is isolated and the faulted circuit is repaired, improving system availability.

10.5.5 Distributed Generation and Demand Response Management (Distributed Energy Resource [DER])

DER system is generation or storage technology [116, 117] that is used to inject energy into the transmission and distribution system or into a local network that is not useful in parallel with the public network. It is distributed geographically according to different factors, including climatic conditions, local energy demand, and land ownership [118].

The implementation of DER offers possibilities to reduce emissions of CO_2, minimize grid losses, save energy, improve grid controllability and monitoring, optimally use the existing grid, and enhance the visibility of DER for transmission system operators [119]. It could be feasible through controlling the reactive energy flow between the DER and the grid.

The DER will operate in both modes connected to the network and on the island. In the network connection mode, it could be used for voltage–ampere reactive management, loss minimization, and peak reduction. In the island mode, DER will be used to deliver power directly to customer loads [102]. Any DER system and its components can be modeled using the functional decomposition principle that is in the following:

- IEC 61850-90-10 – DER scheduling.
- Communication systems IEC 61850-7-420 for DER.
- IEC 61850-90-7: IEC 61850 object models for storage, solar panel, and other DER inverters.
- IEC 61850-90-9: IEC 61850 object model for energy storage system.

The conceptual view of the LNs that could be used for specific types of management systems of DER is indicated in Figure 10.6. These DERs include such technologies as storage, solar panels, wind farm, and virtual power plant.

10.5.5.1 Storage

The energy storage system is utilized to save energy, the same manner of how a rechargeable battery operates. It is numerous sort of energy storage device; however, the technology limits the amount of energy that could surely be stored. It can be used to minimize peak load [120]. In fact, by installing powers systems with a certain capacity in the grid that store the excess power during the night and then release it during the peak load of the day, it effectively improves the use of electrical equipment and the reliability of supply [121], i.e. pumped hydroelectric plants account for most of this deposit at present.

Pumped hydroelectric energy allows the storage of large amounts of energy, and the accumulation of energy with compressed air is a less widespread technology that uses non-peak renewable electricity to compress and store air, which can then

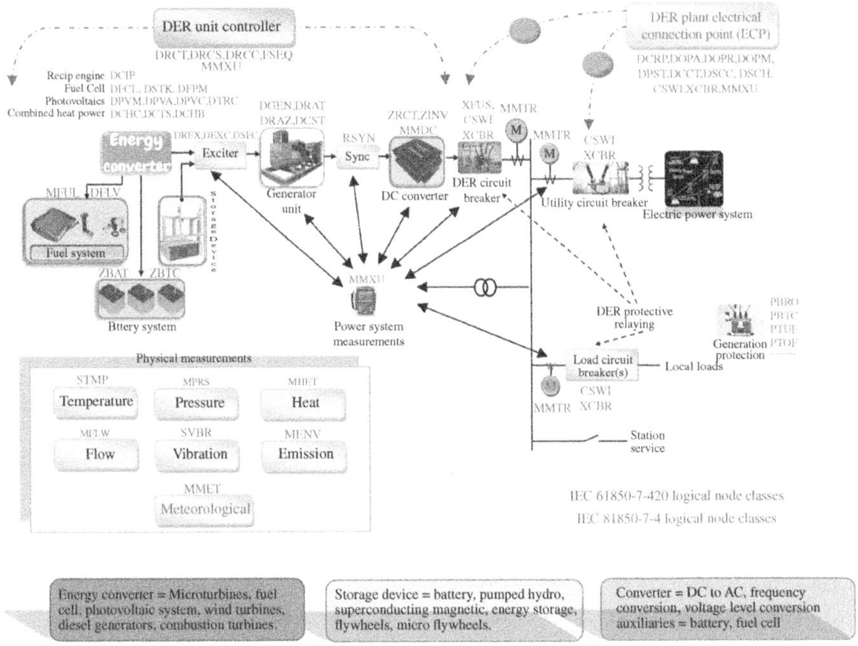

Figure 10.6 Conceptual view of LNS used in different types of DER management system. Source: Reproduced from IEC [126].

be used to recover electricity [122]. The following are some of the energy storage devices used in DER [123]:

- **Uninterruptible power supply (UPS):** this is an electrical device that supplies emergency power to a load when the input power supply fails.
- **Superconducting magnetic energy storage (SMES)**: the SMES system stores electric energy using a superconducting coil cooled to low temperature. This system uses an inverter/converter to convert alternating current (AC) to direct current (DC) or transforms DC back to AC that can be sent to a public network.
- **Flywheel system**: flywheel transforms mechanical or electrical energy into rotational energy and invests it by using it when necessary. Sometimes it is used as an alternative or as a complement to UPS technology for large data facility technologies and other uses. It tends to be more respectful of the environment than battery technology and often has a longer lifespan.
- **Hybrid system**: it is a combination of the previous systems destined for an unusual or specific application.

- **Large battery**: it refers to advance electricity storage technology that has a much higher capacity than a regular UPS system. This technology generally uses electrochemical energy storage method, which maintains a high energy density and is flexible and scalable in particular [102, 114]:

The power plants use lead-acid battery for backup power. For most application, vanadium oxide flow and sodium sulfide batteries are more effective. The accumulation of battery power on a large scale can help peak shaving. The lithium-ion battery finds its applications in larger power demands, portable battery application, and cycling performance that, for example, allow plug-in hybrid electric vehicles (PHEVs) to store distributed power that can reduce fluctuations in charge and generation of manageable load and discharge of energy in the network when it is necessary.

The electrical storage must fulfill a series of features in the network including the rotation reserve function, which serves as a manageable load, load leveling, stabilization of the power system, load reduction, and reactive power support.

10.5.5.2 Solar Panels

The photovoltaic technology is an energy technology destined to convert light directly into electricity. It can range from very small grid-connected PV system at residential customer sites to a very large PV plant. It can also use solar tracking technology to enhance overall system performance with the integration of battery storage as a solution [124, 125]. The photovoltaic inverter can adjust the power factor, and this means that the overall efficiency of the system can be improved if the load data is available and shared through real-time communication. The customer's EMS/SCADA and the energy utilities will manage the power factor (phi) adjustment of the inverters connected to the network by getting value for the phase angle [31].

The object models and data modeling for photovoltaic and different DER inverters are defined and described in the IEC 61850-90 90-7 and IEC 61850-7-420 standards [126]. These features consist of watt/power factor control, fixed VAR, voltage/VAR, fixed power factor, and voltage/watt control of active power, and frequency/watt. They allow the transfer of energy quality data (for example, root mean square (RMS) data, transients, and harmonics data) with a considerable enhancement in the photovoltaic modules and protection of inverters [31, 126, 127]. Table 10.6 demonstrates some roles and description for some LNs in solar panels.

Figure 10.7 demonstrates the physical model of the photovoltaic system, LDs, and LNs. The LNs in blue are the LNs standards defined above in IEC 61850-7-4 of IEC 61850, whereas the LNs in red belong to the new subpart of IEC 61850-7-420. MMET meteorological LN data could be used for photovoltaic and wind DERs for further information; please refer to Ref. [126, 128].

Table 10.6 Roles and descriptions for some logical nodes in PV.

Logical nodes	Roles and descriptions
DPVC, DTRC, PTOC, PHIZ, MMXU, CSWI, DPVC, and DTRC	"Contain the information required to augment the power produced off the array and the tracking controller to follow the movement of the sun"
PTOC and PHIZ	"Detect the overcurrent AC flow and the DC ground fault up to the threshold and send the detection signal to the CSWI node in the case of failure"
CSWI	"Sends the opening or closing signal to the XCBR node in the IED breaker"
STMP and MMET	"Contain the information and settings for temperature measurements and meteorological conditions, respectively"
ZINV	"Defines the configuration and the status information of the device that transforms DC into AC"

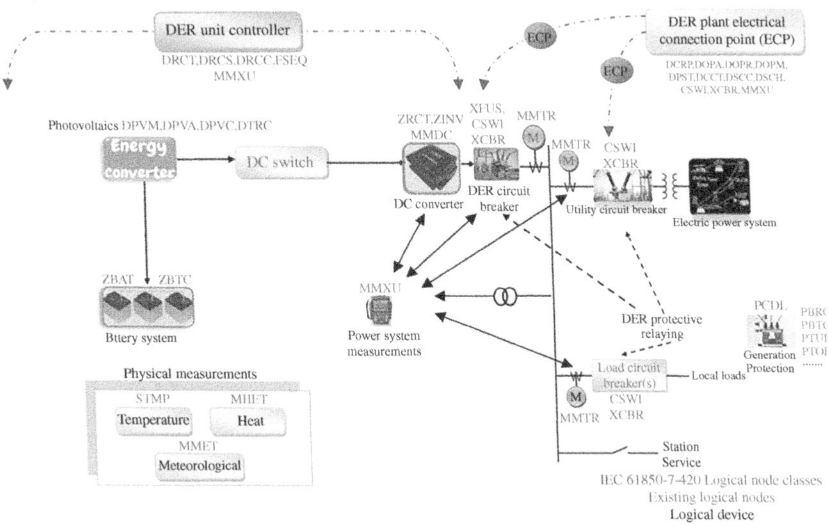

Figure 10.7 Solar panel system LDs and LNs. Source: Reproduced from IEC [182].

10.5.5.3 Wind Farm

A wind farm consists of several individual wind turbines used to produce electric power. The critical requirements for wind farm are to lay out a reliable communication network to monitor the wind farms. The IEC 61400-25 bases on communication for control, monitoring, and wind farm condition monitoring.

There are three domains for the wind farm, depending on intelligent systems [129, 130]:

- **Domain of turbine**: "wind turbine exchanges data with different wind turbines to optimize availability, power generation, and lifetime."
- **Domain of network**: "enables a connection between the principal network and the wind farm. It must be rapid, scalable, reliable, and it needs to fulfill several quality services for other applications."
- **Domain of application**: "supports the control center operators to enhance the principal features of control, monitoring, and analysis."

In effect, IEC 61850 has been expanded to contain new applications, i.e. wind turbine applications "IEC 61400-25." These applications cover all the necessary components for the operation of wind farms including the electrical system, the wind farm management system, and the meteorological system. IEC 61400-25 is object oriented, which permits using virtualized models. It consists of six parts, as listed in Table 10.7.

Table 10.8 describes the LNs used for wind power systems. First, four LNs are for an entire wind farm, and the remaining are designed for the wind turbine. Figure 10.8 illustrates the LNS missions throughout wind farm. Some of them, for example, XCBR, are defined in both IEC 61850 and IEC 61400-25-2 standards.

10.5.5.4 Virtual Power Plant (VPP)

A virtual power plant (VPP) is a terminology used to aggregate DERs to make them appear as an alone or a single and bigger power plant [132, 133]. The virtual power plant can apply IEC 61850 communication in two directions for the low- and medium-voltage substations. This virtual power plant let the coordinated control of the units of DERs, which consecutively enables for a greater percentage of intermittent power generation in the network [42]. Their electronic power converters can supply inductive or capacitive reactive power to maintain the voltage level. Usually, these units are configured to keep the reactive power close to 0 at the DER node [134].

The authors in Refs. [135, 136] have suggested using IEC 61850 for virtual power plant. The Smart Grid Architecture Model (SGAM) splits the power system components into zones and domains (Figure 10.9). The management of the DER system covers the field, process, operation and station zones, as well as the customer

Table 10.7 Definition and description of IEC 61400-25 protocol.

Part	Title	Definition and specification
1	A general summary of models and standards	"Preliminary guide, a general description of the essential specifications and fundamental standards and a model"
2	Data models	"The information labeled with the specific name of the wind farm describes the crucial and common data of the process, the metadata (data on the data, for example, scaling factor or unit conversion), and the configuration data of a wind farm"
3	Data transfer models	"All processes and metadata can be transferred for the corresponding utilities. Accessing to metadata (as well as configuration data regarding the model of the turbine and utilities and the communications stack) presents self-description of a device"
4	"Mapping to communication protocols"	"Web services, IEC 60870-5-101, IEC 60870-5-104, MMS, DNP3, DA, XML, and OPC"
5	Compliance test	"Specific techniques for compliance test of implementation and standards measuring techniques applied for tags or attributes"
6	"Data and LN classes for condition monitoring"	"Data models to be used in the condition monitoring system"

premises and distribution domains. There are also interactions for the market and enterprise zones and potentially for the transmission domain [42, 134–137].

IEC 61850-90-15 will represent a series of services requested by a VPP. Indeed the grid services using IEC 61850 for VPP require the following group of features based on [42, 137]:

(1) **Management of DER system**: link a DER to the VPP controller, read the functionality of the DER service, and update the list of VPP features and resources according to the topology of the grid.
(2) **Control of the DER system**: it includes data acquisition services, control of a DER aggregation, and division and distribution of control requests to the different DER.

Table 10.8 Wind farm systems logical nodes.

Logical nodes	Description	Specification	Mandatory "M" or optional "O"
WALG	"Wind turbine analog log data"	WT	O
WALM	"Wind farm alarm data"	Wind power plant (WPP) Wind turbine (WT)	O
WAPC	"Wind farm active power control data"	WPP	O
WCNV	"Wind turbine converter data"	WT	O
WGEN	"Wind turbine generator data"	WT	M
WMET	"Wind farm meteorological data"	WPP	O
WNAC	"Wind turbine nacelle data"	WT	M
WREP	"Wind turbine report data"	WT	O
WROT	"Wind turbine rotor data"	WT	M
WRPC	"Wind farm reactive power control data"	WPP	O
WSLG	"Wind turbine state log data"	WT	O
WTOW	"Wind turbine tower data"	WT	O
WTRF	"Wind turbine transformer data"	WT	O
WTRM	"Wind turbine transmission data"	WT	O
WTUR	"Wind turbine general information"	WPP WT	M
WYAW	"Wind turbine yawing data"	WT	M

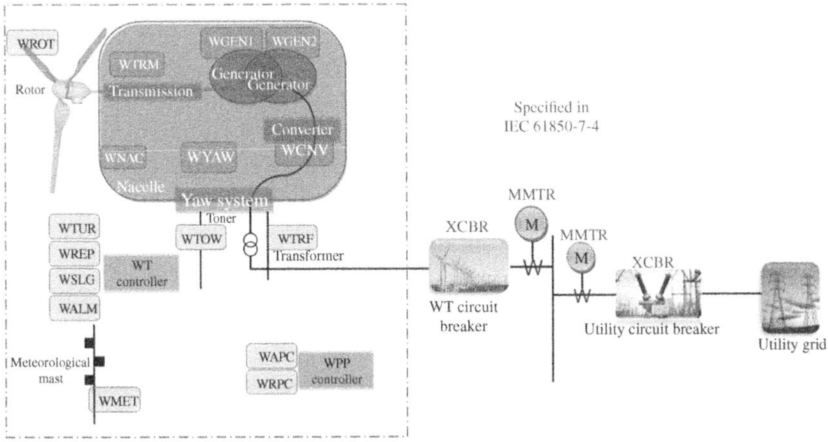

Figure 10.8 IEC 61400-25 wind farm systems logical node distribution. Source: Reproduced from Schwarz [131].

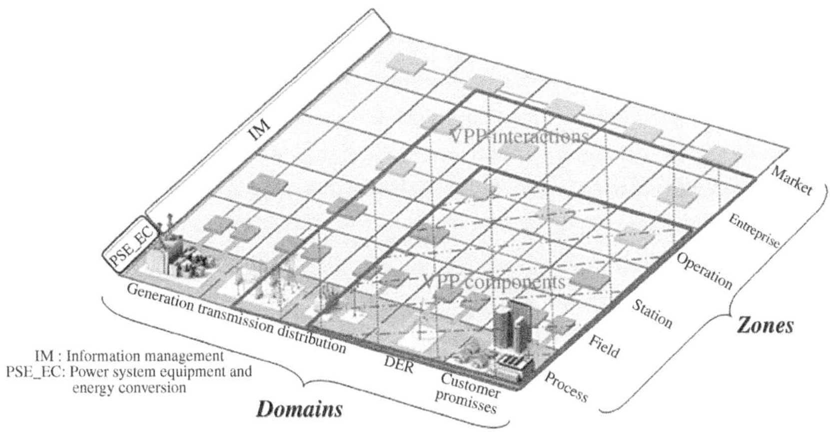

Figure 10.9 Domain and hierarchical zones of SGAM. Source: Reproduced from Guise [154].

(3) **Control of DER or micro-grid**: SCADA of the participating units in the VPP. The peak power (P) and balancing (B) applications require the VPP to compute the individual set points for the active power. The compensation of the reactive power (R) and the loss of the network (L) require that the set points of the reactive power to be computed and sent to the individual VPP units.

10.5.6 Advanced Metering Infrastructure (AMI)

AMI, which is also known "smart metering and infrastructure" (SMI), offers a two-way communication network for intelligent networks [138]. AMI typically make use of a communication interval of 15 minutes to once per hour [139]. Multicasting and broadcasting transmissions help to prevent from sequential meter reading in the AMI system [140]. AMI includes the communication systems, integration of smart meters, and software and hardware that allow the measurement, collection, storage, analysis, and use of energy between public services and smart meters or customers and smart meters [88, 141, 142]. Table 10.9 lists the number of components in an AMI system.

10.5.7 Electric Vehicle (EV)

An EV is an automobile that uses one or more electric propulsion engines. Renewable energy resources, such as wind energy, solar power, or any hybrid energy systems, could power an EV. The electric energy could be stored in the vehicle battery, fuel cell, flywheel, or supercapacitor. When combined with an intelligent network, an EV can behave like a distributed storage device and a load [143]. By being linked to the network while no longer in use, the EV battery can supply power at peak load times and, therefore, enhance the reliability of the network. This technique is known as V2G [144].

By supplying electric energy to the network, this technique offers many advantages, e.g. active/reactive power compensation, frequency and voltage regulation, electric demand-side management, harmonic filtering, spinning reserve, and load balancing [88, 145].

The IEC 61850 [146] allows the regulation of the communication of the substation and its latest extension IEC 61820-7-420 [147] targets at abstract information and data models associated with DER units [148, 149]. Figure 10.10 shows the general EV model with IEC 61850-7-420. The EVs are linked to the network through an inverter and a DC switch. Extended FSEQ, DRCC, DRCS, and DRCT could be used for supplementary network control as in distinct models of DER (see Table 10.10).

These models can be utilized for a single EV and V2G aggregates consisting of several EVs. The common nature of EVCT class makes the suggested model very adaptable. Diverse EVs constructed through distinct companies may be effectively modeled and used in a pool or one by one [148, 150].

Table 10.9 Components of AMI system.

Components of AMI system	Definition and role
"Smart meter"	It is a digital device that measures energy consumption instantaneously, registers, and stores this measurement at intervals. It can carry out several tasks: – Dynamic pricing – Customer usage data for service and consumer – Remote on/off functions – Load shedding or demand management – Communications with intelligent devices within HAN
"Wide area network (WAN)"	It supports communication between the intelligent network and the public services network, which gathers information from various NANs and communicates them to the control center It links networks of small and highly distributed areas that deliver energy systems at distinct locations It can envelop thousands of square kilometers area; therefore data transmission speeds can reach 10^4–10^5 kbit/s
"Home (local) area network (HAN/LAN)"	HAN or LAN is a communication system to transmit the price in real time, the cost of the load, and the control of the load inside the customer's property. Energy management features can include the following: – In-home display for showing energy consumption – Local control actions – Response to price signals It needs to envelop over to 200 m^2 of areas and supports 10–10^2 kbit/s
"Neighborhood area networks (NANs)"	"NANs are used to collect data from a smart meter to transfer energy data and control data among other units. It envelops an area of 2–16 km^2, and data transmission speed is approximately 10–10^3 kbit/s"
"Meter data management system (MDMS)"	"The MDMS is a database with models (mechanisms) that allow interchange with different records systems which include outage management system, billing systems, consumer information system, etc. One of the main features of an MDMS is to carry out estimation, editing, and validation on the AMI data"

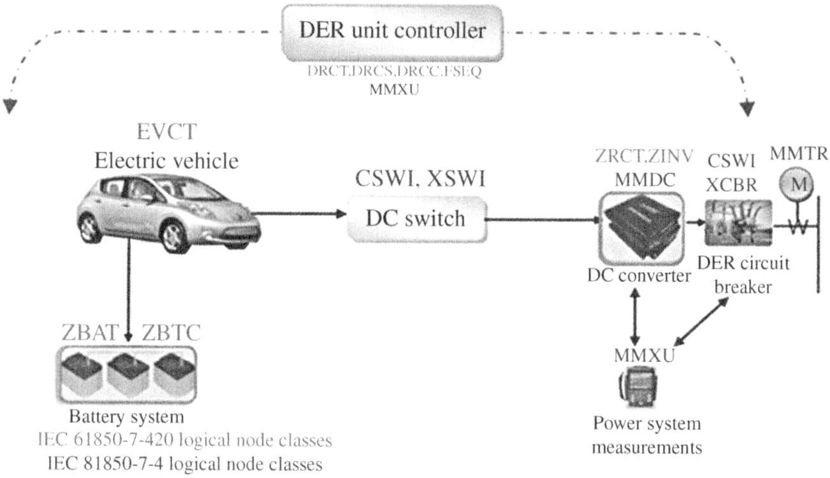

Figure 10.10 EV control class modeled with IEC 61850-7-420.

Table 10.10 EV classes and other DER models.

Logical nodes	Descriptions
EVCT	Electric vehicle control
ZBAT	Battery
ZBTC	Charger
CSWI and XSWI	"Connection to the EV network is made by a DC switch"
ZCRT and ZINV	"Inverter"
EVCT	Electric vehicle control
DRCT	"Controller that supervises of individual DER or many DERs linked to one controller"
DRCS	"Provide data on the state of the controller as its operation mode: automatic/ manual, electrical connection, etc."
DRCC	"Supervisory control on DER and having a complete list of actions that can be performed for the performance goal"
FSEQ	"Provides data on the cycle of operations during shutting down or starting up of a DER"

10.6 Advantages of IEC 61850 (Requirements of Smart Grid IEC 61850)

Many advantages can be achieved using IEC 61850 for the integration of electrical substation, control, and automation. The principal benefits consist of the following.

10.6.1 Communications Bandwidth

Legacy communication protocols are developed by limiting the number of bytes that the protocol has used due to the severe bandwidth constraints typical of serial link technology 10–15 years ago. However, as Ethernet and modern network protocols such as TCP/ IP are widely used, the protocols do not take advantage of the large bandwidth increase that modern network technologies offer by providing a higher level of functionality that can significantly reduce the implementation and operational costs of the automation substation [1].

IEC 61850 allows devices to rapidly communicate status and data using GSSE and GOOSE on the station LAN without having to connect separate connections for each relay. Therefore, it automatically minimizes wiring costs by taking advantage of the bandwidth of the station's LAN for these signals and construction costs by minimizing the need for ducts, conduit, trenching, etc. [1].

Overall, all substations can be accessed through WAN, which means that the bandwidth efficiency of the protocols used is less important than, for example, in powerline carrier technologies. The bandwidth efficiency remains an issue for recovering sensor data over wireless connections or from the energy meter outside the substation, even if the wireless LAN with IEC 61850 applies to distributed automation [42, 151–153].

10.6.2 Interoperability

As mentioned before, the interoperability is recognized as the capacity of two or more IEDs from the same provider, or multi-providers, to transfer information and use them for the correct execution of distinct functions [15, 22]. In other words, how to make different devices and tools from different vendors work together.

Figure 10.11 shows the mapping of VPP communication, information, function, and business requirements to the SGAM. The SGAM also defines five interoperability layers between business objectives and processes, functions, information models, communication protocols, and finally components [42, 154].

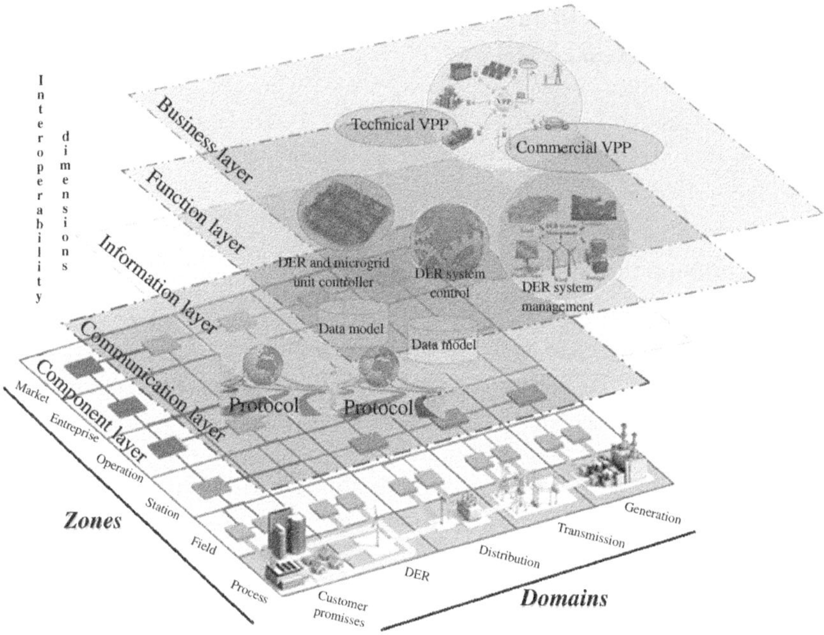

Figure 10.11 Mapping of the VPP. Source: Reproduced from Guise [154].

10.6.3 Cybersecurity

It is a terminology used with the virtual environment, thanks to IEDs, which present automation and control functions. The customer is outside the secured LAN and tries to access the functions of SAS; in this situation, the security of the data is a major concern [31]. Therefore, different control access system should be considered, such as authentication, secure connection, and encryption and prevention of eavesdropping, spoofing, and intrusion detection [155] from malicious activities [156, 157]. Hence, it provides security improvements no longer for MMS but also for SMV and GOOSE messages. IEC 62351 was developed to apply for end-to-end security problems in the communication of electric system [158]:

- **IEC 62351-6**: "Security of communication and data for IEC 61850 peer to peer profiles."
- **IEC 62351-3**: "specifies the communication and the authentication profiles of system messages, as well as TCP/IP."
- **IEC 62351-4**: "defines the strong authentication mechanism to be used with the MMS protocol."

These definitions allow providing integrators and manufacturers the tools indispensable to apply security for GOOSE stack and the IEC 61850 [156]. Although IEC 62351 solves many security problems, issues still remain [159].

10.6.4 Information Security and Privacy

The security of information is the usage of preventing disruption, disclosure, unauthorized access, use, inspection, modification, destruction, or recording of information [160]. It is related to the protection of data saved, transmitted, and processed to fulfill the functions and purposes of the information systems of an organization while the privacy of data refers to the protection of information associated with the identity of a subject. It focused on safeguarding rights of the individual concerning the same data.

The cybernetic threats of the intelligent network, e.g. the Stuxnet worm [161] or viruses, spyware, Trojans, denial of service (DoS), keyloggers, etc., can violate economic stability and national security. Attackers have continually addressed power stations and SCADA system.

The transition from a locked control system to an unlocked IP network introduces a new field of vulnerabilities. As an illustration, authentication and data integrity could be dealt with by network attacks, i.e. identity theft, impersonation, or DoS attack. In the same way, data security can be dealt with by internal attacks and sabotage such as Trojans and viruses [161, 162].

It becomes important taking into consideration the potential openness of the Internet and NANs. Once the vulnerability is identified, it becomes not difficult for the attacker to cascade through the intelligent network. For standardizing this subject, professionals and academics work together and try to establish policies and basic guidelines for the password, cryptographic software, firewalls, antivirus software, legal responsibility, and training standards for users/administrators [163].

10.6.5 Security Aspects of IEC 61850 in Smart Grid Applications

When the IEC 61850 protocol was conceived in 2004 [164], its main objective was to facilitate interoperability of numerous products from various vendors, and the interconnection was mainly isolated within the substations. To achieve such capabilities, the protocol had made use of MMS, GOOSE, and SVs for the communication network. Soon however, the network had undergone further expansion to include connections beyond merely within the substations allowing wider accessibility. Nevertheless, such openness also made way for bigger security issues.

10.6.5.1 How Security Can Be Breached

Since the smart grid make use of cyber network, the security of the power systems can be breached by means of cyber attack that can be dissected into several aspects such as its motivation, vectors, and techniques. While the motivation deals with the reason of the attack, the vectors and techniques are essentially more of our interest here since they are pertaining to the how the attack is being accomplished.

Attack Vectors It constitutes the path or means of accessing the network or a computer by exploiting various elements of the system vulnerabilities. For example, malware can be introduced in the system unknowingly through the computers used during maintenance or any other activities by supplier support team that access the network such as the incident at Siemens [165]. Such malware may possibly be introduced even earlier in any IED or other network equipment during their manufacturing process. When these equipment are installed in the substation, the malware gets direct access to the network. Another possible vector may simply come from any ill-intent individual that has direct physical access to the IEC 61850 network such as the substation personnel, cleaning crews, etc.

Techniques of Attack Once the IEC 61850 network is accessed by any means of the attack vector, many relevant Ethernet techniques of attack can be used at the data link layer (layer 2) of OSI model such as ARP attacks, MAC flooding attacks, spanning tree attacks, multicast brute force attacks, VLAN trunking protocol attacks, private VLAN attacks, identity theft, VLAN hopping attacks, MAC spoofing, and double-encapsulated 802.1Q/nested VLAN attacks [166, 167]. The upper layers of the network may even be unaware of the network security breach [168].

10.6.5.2 Effects on the Security

Any compromised security of the smart grid network could affect the attributes of the network and the power system in terms of its availability, integrity, confidentiality, accountability, and vulnerability [169]. While cybersecurity can be considered a rather mature subject in many applications such as banking systems, the same cannot be said about its implementation in the power system. Unlike those systems that purely consist of cyber system, the electrical grid is actually a cyber-physical system. Hence, any error or attack on the data interchange could possibly lead to grave physical consequences [170]. For example, consider the implication inflicted by any alteration of vital data values in GOOSE message. Since GOOSE communicates critical information such as alarm, status, and control, any value changes could cause breakdown in automation that can damage field devices like circuit breakers or power transformers. More severely, the fault can cascade failures and power outage to a part of or even the whole cities, states, or regions [167].

10.6.5.3 Efforts to Address the Security Issues

Concerns over security breaches on critical infrastructure (CI) had grown over the years especially post-11 September 2001 incidents. Several establishments had proposed various initiatives in the efforts to ensure the grid security. For example, the North American Electric Reliability Corporation (NERC) produced a set of critical infrastructure protection (CIP) cybersecurity standards to tackle potential risk of cyber and physical attacks that might compromise the power grid integrity [171], and National Institute for Standards and Technology (NIST) released the "Guide to Industrial Control Systems (ICS) Security" that identifies threats and susceptibilities and also prescribes the needed security measures for a cyber network [172]. Nevertheless, more works are still needed to find robust and effective solutions for the smart grid security [170, 173].

Classical Approaches In ensuring cyber networks security, various methods had been well implemented into layers of protection. First of all, physical security can be emplaced to prevent attack vectors. Failing, which measures at the Ethernet layer 2, can be taken such as making use of virtual local area networks (VLANs) to isolate the communication links to specific equipment and employing software-defined networking (SDN) for better network management [166, 174].

However, these traditional IT techniques would not be effective against attack vectors that have direct access to the facilities. Essentially, more rigorous security measures are needed. Such methods include encryption of data packets or addition of digital signature to prevent monitoring by possible attacker and authentication to thwart spoofed packets. Alternatively, a monitoring system that comprehends the IEC 61850 protocol and is able to check GOOSE message content can be deployed to detect suspicious and inconsistent messages [167].

IEC 62351 To improve the security of data interchange, the IEC62351 standard was released in 2007 with the intention to authenticate data transfers using digital signatures, to detect unauthorized intrusion, and to deny spoofing or eavesdropping [175]. The intention was to boost the security of IEC 61850 protocol including MMS, GOOSE, and SV messages. Although IEC 62351 solved many issues, some difficulties still persist.

The Issue of Latency Certain message protocol such as GOOSE are designed to cater for time-critical operations that timing requirement proved to be difficult to comply when appended with known approaches employed to enhance security in other regular cyber systems [170, 173]. One of the main challenges in implementing security measures for communications between IEDs is due to latency.

For example, GOOSE messages for class P1 type 1A (breaker trip functions) allocate only 4ms maximum delay [176]. Therefore, any security measures such as encryption or others resulting in more than 4 ms delay or latency will not meet the specification of the IEC 61850.

Even in the IEC 62351, it is stated that "for applications using GOOSE and IEC 61850-9-2 and requiring 4 ms response times, multicast configurations and low CPU overhead, encryption is not recommended" [175].

Therefore, in facing this conflict between meeting quick reaction and security requirement, the tendency of some manufacturers is to forsake the security mechanism while in favor of the processing speed.

Encryption and Authentication in Synchrophasor Technology Synchrophasors are values of electrical quantities measured by PMU placed at various parts of the grid. Then, these data are sent to a phasor data concentrator through WAMS [177]. Initially, the IEEE C37.118 standard was used in supporting both serial and IP-based multicast transmission of the synchrophasors. However, it did not stipulate any mechanism to prevent security breaches. The need for security enhancement spurs the development of IEC 61850-90-5 standard, which prescribes encryption and authentication as security measures. While the IEC 61850 standard is mainly used for substation automation, IEC 61850-90-5 supports the PMU measurement transmissions across a WAN [178]. IEC 61850-90-5 essentially maps PMU data formats into the IEC 61850 protocols to enable communications with PDC and other PMU devices.

To transmit GOOSE and SV messages over WAN, IEC 61850-90-5 specifies two approaches: either to tunnel the GOOSE and SV protocols over high-speed communication networks or to extend the GOOSE and SV messages as R-GOOSE and R-SV. R-GOOSE and R-SV messages are the preferred method as they can be multicast in WANs, whereas tunnel requires dedicated gateways and only supports point-to-point message exchanges in WAN. The results conducted in [59] show that the authentication process and additional encryption of R-GOOSE/R-SV only introduced negligible computational delays.

10.6.6 Reliability of Technology

Reliability is the probability that the object will carry out a required function without failing under the conditions declared for a certain period.

Usual reliability measurements for the general system would be as follows [179, 180]:

- Failures per several operations
- Mean downtime (MD)
- Percentage of availability
- Meantime between failures (MTBFs)
- Meantime to failure (MTTF)

Even though the security improvement for IEC61850 [112], which is IEC62351, has been released as a technical standard, this standard entirely provides authentication of data exchanges by way of encryption and digital signatures. The message retransmission system is required to perform a one-to-many transmission and to allow the subscriber to realize that the communication channel is in good condition. However, depending on the choice of the final stable retransmission time, it may not be enough to assure the reliability of mission-critical activities [181]. To enhance the reliability of GOOSE messages, different duplicates of the original GOOSE message must be published when the state changes. This system is used to ensure that the subscriber IED can receive a copy of the originally published GOOSE messages and function correctly. A VLAN is one of the technical solutions that reduce the impact of the diffusion/multicast phenomenon in LAN. It can be carried out by way of originating a necessary subnet by splitting the physical network connection and reducing the traffic flow by restricting it to a single VLAN [81].

10.7 Conclusion and Perspective

This chapter has discussed a detailed investigation and application of the IEC 61850 in the smart grid. The IEC 61850 covers several features and technologies included in topics such as AMI, DMS, EMS, and EV. This chapter also explained in details how IEC 61850 could be implemented over DERs, i.e. storage, solar panels, wind farm, and virtual power plant. IEC 61850 does not alone support any IT security measures by itself. Multi-vendors must also ensure that their IEDs meet the requirements of IEC 62351. Finally, the work in this chapter suggests guidelines to improve the use of IEC 61850.

During our analysis and consultation of the literature, no one of authors talked about the drawbacks of IEC 61850 for smart grid application that can be resumed as follows:

- There are not existing LNs for some applications such as AMI, VPP, and automated FLISR.
- As for the interoperability, the problems faced by the system designers in integrating IEC 61850 devices from different vendors are due to the different SCL configuration procedures and the GOOSE and SV configuration. In other words, there is not an unified SCL configuration and also GOOSE and SV configuration.

The power system has usually worked with AC, although today the majority devices work with DC inside and a good number of DERs produce electricity in DC such as photovoltaic, EV, etc. Furthermore, storage systems such as battery and fuel cell or hybrid system have a DC nature.

DC grid has the capability to make smart grid applications in a more easy way. With DC grid, DC source and DC load could be directly. DC voltage could improve the stability of the grid and enhance its reliability. From all these reasons, our future work will be devoted to study of the application of IEC 61850 for multi-terminal DC grid automation.

Acknowledgments

"The authors thank the International Electrotechnical Commission (IEC) for permission to reproduce Information from its International publications. All such extracts are copyright of IEC, Geneva, Switzerland. All rights reserved. Further information on the IEC is available from www.iec.ch. IEC has no responsibility for the placement and context in which the author reproduces the extracts and contents, nor is IEC in any way responsible for the other content or accuracy therein." The authors would also like to thank CEN CENELEC Group, Ralph Mackiewicz from SISCO, Inc. (USA), and Karlheinz Schwarz from Schwarz Consulting Company (SCC), Germany, for their permissions to use their materials (graphs and figures).

References

1 Mackiewicz, R. (2006). Overview of IEC 61850 and benefits. In: *2006 IEEE PES Power Systems Conference & Exposition, Atlanta, GA, USA*, 623–630.
2 Adamiak, M., Baigent, D., and Mackiewicz, R. (2009). IEC 61850 communication networks and systems in substations: An overview for users. *Protection & Control Journal*: 61–68.
3 Sidhu, T., Kanabar, M., and Parikh, P. (2008). Implementation issues with IEC 61850 based substation automation systems. In: *5th National Power Systems Conference (NPSC)*, 473–478. India: IIT Bombay.
4 Gao, H., Jin, W., and Liu, G. (2008). Simulation study on the delay of end-to-end data communication for protective relaying in substations. *Frontiers of Electrical and Electronic Engineering in China* 3 (2): 246–250. https://doi.org/10.1007/s11460-008-0044-0.
5 Ali, H. and Dasgupta, D. (2012). Effects of time delays in the electric power grid. In: *International Conference on Critical Infrastructure Protection ICCIP 2012: Critical Infrastructure Protection*, vol. VI, 139–154.
6 Rush, P. (2002). *Alstom NPAG Network Protection & Automation Guide*, 497. Alstom T&D.

7 Brunner, C., (2008). IEC 61850 for power system communication. *2008 IEEE/PES Transmission and Distribution Conference and Exposition*. Chicago, IL, USA, 1–6.

8 Amin, S.M. and Wollenberg, B. (2005). Toward a smart grid: power delivery for the 21st century. *IEEE Power and Energy Magazine* 3 (5): 34–41. https://doi.org/10.1109/mpae.2005.1507024.

9 Amin, S.M. (2010). Smart grid: toward stronger, smarter, and more secure energy infrastructure. In: *MN Senate Energy*, 1–87. Minnesota, USA: Utilities, Technology & Communications Committee.

10 Phuangpornpitak, N. and Tia, S. (2013). Opportunities and challenges of integrating renewable energy in smart grid system. *Energy Procedia* 34: 282–290. https://doi.org/10.1016/j.egypro.2013.06.756.

11 Mohamed, M., Eltamaly, A., Farh, H., and Alolah, A. (2015). Energy management and renewable energy integration in smart grid system. In: *2015 IEEE International Conference on Smart Energy Grid Engineering, Oshawa, ON, Canada*, 1–6.

12 Speer et al. (2015). *Role of Smart Grids in Integrating Renewable Energy*. United States: National Renewable Energy Lab. (NREL).

13 Yu, F.R., Zhang, X., and Leung, V. (2013). *Green communication and networking*, 1ste, 399. CRC Press.

14 Schwarz, K. (2005). IEC 61850 also outside the substation for the whole electrical power system. In: *15th Power Systems Computation Conference PSCC, Liège, Belgium*, 1–7.

15 IEC TS 61850-2:2003 (2003). | IEC Webstore | LVDC, Webstore.iec.ch, https://webstore.iec.ch/publication/6009.

16 Xu, T., Hou, H., Yu, H. et al. (2007). Analysis on IEC 61850 interoperability support. In: *2007 IEEE Power Engineering Society General Meeting, Tampa, FL, USA*, 1–6.

17 Cheng, X., Lee, W., and Pan, X. (2017). Modernizing substation automation systems: adopting IEC Standard 61850 for modeling and communication. *IEEE Industry Applications Magazine* 23 (1): 42–49. https://doi.org/10.1109/mias.2016.2600732.

18 IEC 61850-10:2012 (2012). | IEC Webstore | LVDC, Webstore.iec.ch, https://webstore.iec.ch/publication/6008.

19 IEC (2003). *Communication networks and systems in substations*, Parts 1–10. Geneva: IEC.

20 Buchholz, B. and Styczynski, Z. (2014). *Smart Grids –Fundamentals and Technologies in Electricity Networks*, 1ste, 396. Berlin, Heidelberg: Springer, Imprint/Springer Vieweg.

21 Kim, C., Choi, M., Ju, S. et al. (2012). Security data extraction from IEC 61850 ACSI models for network and system management. In: *Information*

Security Applications. WISA 2011, Lecture Notes in Computer Science (eds. S. Jung and M. Yung), 142–152. Berlin, Heidelberg: Springer.

22 Golshani, M., Taylor, G., and Pisica, I. (2014). Simulation of power system substation communications architecture based on IEC 61850 standard. In: *2014 49th International Universities Power Engineering Conference, Cluj-Napoca, Romania*, 1–6.

23 Hou, D. and Dolezilek, D. (2010). IEC 61850 – what it can and cannot offer to traditional protection schemes. *SEL Journal of Reliable Power* 1 (2): 1–12.

24 IEC 61850-7-4:2003 (2003) | IEC Webstore | LVDC, Webstore.iec.ch, https://webstore.iec.ch/publication/20080.

25 IEC 61850-7-3:2003 (2003) | IEC Webstore | LVDC, Webstore.iec.ch, https://webstore.iec.ch/publication/20079.

26 Schwarz, K. (2020). News on IEC 61850 and related Standards. How many and which information models are defined in IEC 61850?. http://blog.nettedautomation.com/2020/02/

27 Kirrmann, H. (2012). *Introduction to the IEC 61850electrical utility standard*. Corporate Research: ABB Switzerland Ltd.

28 Midence, R., Davila, H., and Zhan, N. (2013). Interoperability between recorders and relays. In: *Décimo Quinto Encuentro RegionalIbero-Americano del CIGRÉ Foz de Iguazú*, 1–8. Brazil: Cigre ERIAC.

29 Khaled, O., Marín, A., Almenares, F. et al. (2016). Analysis of secure TCP/IP profile in 61850 based substation automation system for smart grids. *International Journal of Distributed Sensor Networks* 12 (4): 11. https://doi.org/10.1155/2016/5793183.

30 IEC 61850-5:2003 (2003). | IEC Webstore | LVDC, Webstore.iec.ch, https://webstore.iec.ch/publication/20075.

31 Gaviano, A., Weber, K., and Dirmeier, C. (2012). Challenges and integration of PV and wind energy facilities from a smart grid point of view. *Energy Procedia* 25: 118–125. https://doi.org/10.1016/j.egypro.2012.07.016.

32 M. Mekkanen, 2015. On reliability and performance analyses of IEC 61850 for Digital SAS, Ph.D. thesis. University of Vaasa, Finland,

33 IEC (2009). Communication networks & systems for power utility automation –Part 6: Configuration description language for communication in electrical substations related to IEDs Edition 2.0.

34 IEC (2013). Communication networks & systems for power utility automation – part 1: Introduction & overview, ed2.0. Technical report, I.E.C, CH-1211 Geneva 20 Switzerland.

35 Liu, N., Chen, J., Luo, H., and Liu, W. (2011). A preliminary communication model of smart meter based on IEC 61850. In: *2011 Asia-Pacific Power & Energy Engineering Conference, Wuhan, China*, 1–4.

36 Hayati, H., Ahadi, A., and Miryousefi Aval, S. (2015). New concept and procedure for reliability assessment of an IEC 61850 based substation and distribution automation considering secondary device faults. *Frontiers in Energy* 9 (4): 387–398. https://doi.org/10.1007/s11708-015-0382-6.

37 Padilla, E. (2015). *Substation Automation Systems: Design & Implementation*, 1ste, 304. John Wiley & Sons Ltd.

38 International Standard (2013). IEC TR 61850-1:2013 I IEC Webstore I LVDC. Webstoreiecch.

39 Samitier, C. (2017). *Utility Communication Networks & Services*, 1ste. Springer International Publishing.

40 Personnel communication "Mackiewicz R", SISCO, Inc. 6605 19 1/2 Mile Road. Sterling Heights, MI 48314-1408 USA. htttp://www.sisconet.com.

41 Lee, H., Kim, G., Lee, J., and Kim, B. (2008). Real-time communications on IEC 61850 process bus based distributed sampled measured values applications in merging unit. In: *Advanced Intelligent Computing Theories and Applications. With Aspects of Theoretical and Methodological Issues* (eds. D. Huang, D. Wunsch II, D. Levine and K. Jo), 1250–1257. Berlin, Heidelberg: Springer.

42 Etherden, N., Vyatkin, V., and Bollen, M. (2016). Virtual power plant for grid services using IEC 61850. *IEEE Transactions on Industrial Informatics* 12 (1): 437–447. https://doi.org/10.1109/tii.2015.2414354.

43 Becker, F., Nohe, S., and Echeverria, A. (2015). Designing non-deterministic PAC systems to meet deterministic requirements. *Pacworld Protection Automation & Control* 2015: 1–4.

44 Hoyos, J., Dehus, M., and Brown, T. (2019). Exploiting the GOOSE protocol: A practical attack on cyber-infrastructure. In: *2012 IEEE Globecom Workshops, CA, United states*, 1508–1513.

45 Jeschke, S., Brecher, C., Song, H., and Rawat, D. (2017). *Industrial Internet of Things*, 1ste, 715. Springer International Publishing.

46 Wen, J., Hammond, C., and Udren, E. (2012). Wide-area Ethernet network configuration for system protection messaging. In: *2012 65th Annual Conference for Protective Relay Engineers, College Station, TX, USA*, 52–72.

47 El Hariri, M., Youssef, T., and Mohammed, O. (2016). On the implementation of the IEC 61850 Standard: will different manufacturer devices behave similarly under identical conditions? *Electronics* 5 (4): 85. https://doi.org/10.3390/electronics5040085.

48 Gholizadeh, N. (2016). IEC 61850 standard & its capabilities in protection systems. Master thesis. Politecnico Di Milano, Italy.

49 Zhao, P. (2012). IEC 61850-9-2 Process Bus Communication Interface for Light Weight Merging Unit Testing Environment. Master thesis. Stockholm, Sweden.

50 Ozansoy, C., Zayegh, A., and Kalam, A. (2008). Time synchronisation in a IEC 61850 based substation automation system. In: *2008 Australasian Universities Power Engineering Conference, Sydney,Australia*, 1–7.

51 Arnold, G. (2011). Challenges and opportunities in smart grid: a position article. *Proceedings of the IEEE* 99 (6): 922–927. https://doi.org/10.1109/jproc.2011.2125930.

52 Lee, J., Lee, S., Bae, J., and Kwon, D. (2013). The PMU interface using IEC 61850. In: *2013 International Conference on ICT Convergence, Jeju, South Korea*, 1125–1128.

53 Xu, Y., Zhang, D., and Yu, Y. (2011). The research of synchronized phasor measurement units in smart substations. In: *2011 International Conference on Advanced Power System Automation and Protection, Beijing, China*, 2262–2267.

54 Gaudi, A. et al. (2014). *Unified Field Area Network Architecture for Distribution Automation*. CISCO.

55 Apostolov, A. (2017). R-GOOSE: what it is and its application in distribution automation? *CIRED - Open Access Proceedings Journal* 2017 (1): 1438–1441. https://doi.org/10.1049/oap-cired.2017.0494.

56 Youssef, T., Esfahani, M., and Mohammed, O. (2020). Data-centric communication framework for multicast IEC 61850 routable GOOSE messages over the WAN in modern power systems. *Applied Sciences* 10 (3): 848. https://doi.org/10.3390/app10030848.

57 Firouzi, S., Vanfretti, L., Ruiz-Alvarez, A. et al. (2016). An IEC 61850-90-5 gateway for IEEE C37.118.2 synchrophasor data transfer. In: *2016 IEEE Power and Energy Society General Meeting, Boston, USA*, 1–5.

58 Madani, V., Picard, S., Yin, Y., and Adamiak, M. (2015). Challenges and lessons learned from commissioning an IEC 61850-90-5 based synchrophasor system. In: *2015 68th Annual Conference for Protective Relay Engineers, College Station, TX, USA*, 842–849.

59 Ustun, T., Farooq, S., and Hussain, S. (2020). Implementing secure routable GOOSE and SV messages based on IEC 61850-90-5. *IEEE Access* 8: 26162–26171. https://doi.org/10.1109/access.2020.2971011.

60 Zima, M., Larsson, M., Korba, P. et al. (2005). Design aspects for wide-area monitoring and control systems. *Proceedings of the IEEE* 93 (5): 980–996. https://doi.org/10.1109/jproc.2005.846336.

61 Zhu, K., Chenine, M., and Nordstrom, L. (2011). ICT architecture impact on wide area monitoring and control systems' reliability. *IEEE Transactions on Power Delivery* 26 (4): 2801–2808. https://doi.org/10.1109/tpwrd.2011.2160879.

62 Korba, P. and Uhlen, K. (2010). Wide-area monitoring of electromechanical oscillations in the Nordic power system: practical experience. *IET Generation Transmission and Distribution* 4 (10): 1116. https://doi.org/10.1049/iet-gtd.2009.0487.

63 Kundur, P., Balu, N., and Lauby, M. (2009). *Power System Stability & Control*, 1176. New York: McGraw-Hill.

64 Gomez-Exposito, A., Abur, A., de la Villa Jaen, A., and Gomez-Quiles, C. (2011). A multilevel state estimation paradigm for smart grids. *Proceedings of the IEEE* 99 (6): 952–976. https://doi.org/10.1109/jproc.2011.2107490.

65 Chakhchoukh, Y., Vittal, V., and Heydt, G. (2014). PMU based state estimation by integrating correlation. *IEEE Transactions on Power Systems* 29 (2): 617–626. https://doi.org/10.1109/tpwrs.2013.2284560.

66 O'Brien, J. et al. (2014). Use of synchrophasor measurements in protective relaying applications. In: *2014 67th Annual Conference for Protective Relay Engineers, College Station, TX, USA*, 23–29.

67 Khorashadi-Zadeh, H. and Li, Z. (2011). Design of a novel phasor measurement unit-based transmission line auto reclosing scheme. *IET Generation Transmission and Distribution* 5 (8): 806. https://doi.org/10.1049/iet-gtd.2010.0744.

68 Nayak, P., Pradhan, A., and Bajpai, P. (2014). Wide-area measurement-based backup protection for power network with series compensation. *IEEE Transactions on Power Delivery* 29 (4): 1970–1977. https://doi.org/10.1109/tpwrd.2013.2294183.

69 Brahma, S. (2006). New fault-location method for a single multiterminal transmission line using synchronized phasor measurements. *IEEE Transactions on Power Delivery* 21 (3): 1148–1153. https://doi.org/10.1109/tpwrd.2005.860273.

70 Izykowski, J., Rosolowski, E., Saha, M. et al. (2007). A fault-location method for application with current differential relays of three-terminal lines. *IEEE Transactions on Power Delivery* 22 (4): 2099–2107. https://doi.org/10.1109/tpwrd.2007.905544.

71 Kosterev, D. (2004). Hydro turbine-governor model validation in pacific northwest. *IEEE Transactions on Power Systems* 19 (2): 1144–1149. https://doi.org/10.1109/tpwrs.2003.821464.

72 Kennedy, J., Fox, B., Littler, T., and Flynn, D. (2011). Validation of fixed speed induction generator models for inertial response using wind farm measurements. *IEEE Transactions on Power Systems* 26 (3): 1454–1461. https://doi.org/10.1109/tpwrs.2010.2081385.

73 Carta, A., Locci, N., and Muscas, C. (2009). A PMU for the measurement of synchronized harmonic phasors in three-phase distribution networks. *IEEE Transactions on Instrumentation and Measurement* 58 (10): 3723–3730. https://doi.org/10.1109/tim.2009.2019319.

74 Melo, I., Pereira, J., Variz, A., and Garcia, P. (2017). Harmonic state estimation for distribution networks using phasor measurement units. *Electric Power Systems Research* 147: 133–144. https://doi.org/10.1016/j.epsr.2017.02.027.

75 IEEE Std. 1547.2-2008 (2008). IEEE Application Guide for IEEE Standard for Interconnecting Distributed Resources with Electric Power Systems.

76 Borghetti, A., Nucci, C., Paolone, M. et al. (2010). Synchronized phasors monitoring during the islanding maneuver of an active distribution network. In: *2010 Innovative Smart Grid Technologies, Gothenburg, Sweden*, 1–8.

77 Best, R., Morrow, D., Laverty, D., and Crossley, P. (2011). Techniques for multiple-set synchronous islanding control. *IEEE Transactions on Smart Grid* 2 (1): 60–67. https://doi.org/10.1109/tsg.2010.2100833.

78 Veichtlbauer, A., Parfant, M., Langthaler, O. et al. (2016). Evaluating XMPP communication in IEC 61499-based distributed energy applications. In: *2016 IEEE 21st International Conference on Emerging Technologies & Factory Automation, Berlin, Germany*, 1–8.

79 Mackiewicz, R. (2006). The benefits of standardized web services based on the IEC 61970 generic interface definition for electric utility control center application integration. In: *2006 IEEE PES Power Systems Conference and Exposition, GA, United States*, 491–494.

80 Zhu, L., Shi, D., and Duan, X. (2011). Standard function blocks for flexible IED in IEC 61850-based substation automation. *IEEE Transactions on Power Delivery* 26 (2): 1101–1110. https://doi.org/10.1109/tpwrd.2010.2091154.

81 Thomas, M. and Ali, I. (2010). Reliable, fast, and deterministic substation communication network architecture and its performance simulation. *IEEE Transactions on Power Delivery* 25 (4): 2364–2370. https://doi.org/10.1109/tpwrd.2010.2042824.

82 Ozansoy, C., Zayegh, A., and Kalam, A. (2009). The application-view model of the international Standard IEC 61850. *IEEE Transactions on Power Delivery* 24 (3): 1132–1139. https://doi.org/10.1109/tpwrd.2008.2005657.

83 Falahati, B., Mousavi, M., and Vakilian, M. (2011). Latency considerations in IEC 61850-enabled substation automation systems. In: *2011 IEEE Power and Energy Society General Meeting, CA, United States*, 1–8.

84 Janssen, M. and Apostolov, A. (2008). IEC 61850 impact on substation design. In: *2008 IEEE/PES Transmission and Distribution Conference and Exposition, Chicago, United States*, 1–7.

85 Holbach, J. et al. (2007). Status on the first IEC61850 based protection and control, multi-vendor project in the United States. In: *2007 60th Annual Conference for Protective Relay Engineers, College Station, TX, USA*, 283–306.

86 Yenginer, H., Cetiz, C., and Dursun, E. (2015). A review of energy management systems for smart grids. In: *2015 3rd International Istanbul Smart Grid Congress & Fair, Istanbul, Turkey*, 1–4.

87 Katiraei, F., Iravani, R., Hatziargyriou, N., and Dimeas, A. (2008). Microgrids management. *IEEE Power and Energy Magazine* 6 (3): 54–65. https://doi.org/10.1109/mpe.2008.918702.

88 Su, W. and Wang, J. (2012). Energy management systems in microgrid operations. *The Electricity Journal* 25 (8): 45–60. https://doi.org/10.1016/j.tej.2012.09.010.

89 Muyeen, S. and Rahman, S. (2017). *Communication, Control and Security Challenges for the Smart Grid*, 576. Stevenhage: The Institution of Engineering & Technology.

90 Li, Y. and Nejabatkhah, F. (2014). Overview of control, integration and energy management of microgrids. *Journal of Modern Power Systems and Clean Energy* 2 (3): 212–222. https://doi.org/10.1007/s40565-014-0063-1.

91 Olivares, D., Canizares, C., and Kazerani, M. (2014). A centralized energy management system for isolated microgrids. *IEEE Transactions on Smart Grid* 5 (4): 1864–1875. https://doi.org/10.1109/tsg.2013.2294187.

92 Ancillotti, E., Bruno, R., and Conti, M. (2013). The role of communication systems in smart grids: Architectures, technical solutions and research challenges. *Computer Communications* 36 (17–18): 1665–1697. https://doi.org/10.1016/j.comcom.2013.09.004.

93 Tsikalakis, A. and Hatziargyriou, N. (2008). Centralized control for optimizing microgrids operation. *IEEE Transactions on Energy Conversion* 23 (1): 241–248. https://doi.org/10.1109/tec.2007.914686.

94 Meng, L., Sanseverino, E., Luna, A. et al. (2016). Microgrid supervisory controllers and energy management systems: A literature review. *Renewable and Sustainable Energy Reviews* 60: 1263–1273. https://doi.org/10.1016/j.rser.2016.03.003.

95 Papadopoulou, E. (2012). *Energy Management in Buildings Using Photovoltaics*, 1ste, 130. London: Springer Verlag.

96 Lehpamer, H. (2016). *Introduction to Power Utility Communications*, 271. Artech House.

97 Mesmaeker, I., Brand, K. and Brunner, C. (2005). How to use IEC 61850 in protection and automation. SCB5, N° 222-ELECTRA 11.

98 De Mesmaeker, I., Rytoft, C., and Reinhardt, P. (2007). Protection and substation automation systems: standardization, integration and information technology. In: *2007 IEEE Power Engineering Society Conference & Exposition in Africa – Power Africa, Johannesburg, South Africa*, 1–6.

99 Higgins, N., Vyatkin, V., Nair, N., and Schwarz, K. (2011). Distributed power system automation with IEC 61850, IEC 61499, and intelligent control. *IEEE Transactions on Systems, Man, and Cybernetics, Part C* 41 (1): 81–92. https://doi.org/10.1109/tsmcc.2010.2046322.

100 Yusa, H. et al. (2011). Application of IEC 61850 to the communication of a large-scale photovoltaic power generation operation - Web service and conformance test. In: *2011 IEEE International Conference on Smart Grid Communications, Brussels, Belgium*, 522–527.

101 Higgins, N., Vyatkin, V., Nair, N., and Schwarz, K. (2008). Concept for intelligent distributed power system automation with IEC 61850 and IEC 61499. In: *2008 IEEE International Conference on Systems, Man & Cybernetics, Singapore*, 36–41.

102 Sorebo, G. and Echols, M. (2011). *Smart Grid Security: An End-to-End View of Security in the New Electrical Grid*, 1ste, 328. CRC Press.

103 Kwon, S., Chu, C., Cho, J., and Yoon, S. (2012). Development of IEC 61850 based Feeder IEDs for self-healing operation in distribution network. In: *CIRED 2012 Workshop: Integration of Renewables into the Distribution Grid, Lisbon, Portugal*, 1–4.

104 Ali, I. and Hussain, S. (2017). Control and management of distribution system with integrated DERs via IEC 61850 based communication. *Engineering Science & Technology, an International Journal* 20 (3): 956–964. https://doi.org/10.1016/j.jestch.2016.11.017.

105 Senke, N. et al. (2012). Application of the IEC 61850 to communication in distribution automation and building energy management systems-evaluation of the applicability of standard logical nodes and data objects. In: *2012 IEEE 3rd International Conference on Smart Grid Communications, Tainan, Taiwan*, 454–459.

106 Giri, J. (2015). Proactive management of the future grid. *IEEE Power and Energy Technology Systems Journal* 2 (2): 43–52. https://doi.org/10.1109/jpets.2015.2408212.

107 Mohagheghi, S. et al. (2011). Applications of IEC 61850 in distribution automation. In: *2011 IEEE/PES Power Systems Conference and Exposition, Phoenix, AZ, USA*, 1–9.

108 Han, G., Xu, B., Fan, K., and Lv, G. (2014). An open communication architecture for distribution automation based on IEC 61850. *International Journal of Electrical Power & Energy Systems* 54: 315–324. https://doi.org/10.1016/j.ijepes.2013.07.013.

109 Zhu, Z., Xu, B., Brunner, C. et al. (2017). IEC 61850 configuration solution to distributed intelligence in distribution grid automation. *Energies* 10 (4): 528. https://doi.org/10.3390/en10040528.

110 Yip, T., Xu, B., Zhu, Z. et al. (2018). Application of IEC 61850 for distribution network automation with distributed control. *The Journal of Engineering* 2018 (15): 993–996. https://doi.org/10.1049/joe.2018.0182.

111 Zhabelova, G. and Vyatkin, V. (2012). Multiagent smart grid automation architecture based on IEC 61850/61499 intelligent logical nodes. *IEEE Transactions on Industrial Electronics* 59 (5): 2351–2362. https://doi.org/10.1109/tie.2011.2167891.

112 Kanabar, M. and Sidhu, T. (2009). Reliability and availability analysis of IEC 61850 based substation communication architectures. In: *2009 IEEE Power & Energy Society General Meeting, Calgary, AB, Canada*, 1–8.

113 Tobar, A., Ul Banna, H., and Cosmin Koch-Ciobotaru, C. (2014). Scope of electrical distribution system architecture considering the integration of renewable energy in large and small scale. In: *IEEE PES Innovative Smart Grid Technologies, Europe, Istanbul, Turkey*, 1–7.

114 SMB Smart Grid Strategic Group (2010). *IEC Smart Grid Standardization Roadmap*, 1ste, 136. IEC.

115 Gill, H. (2010). Smart grid distribution automation for public power. In: *IEEE PES T&D 2010, LA, USA*, 1–4.

116 McDonald, J., Wojszczyk, B., Flynn, B., and Voloh, I. (2012). Distribution systems, substations, and integration of distributed generation. In: *Electrical Transmission Systems and Smart Grids* (ed. M. Begovic), 7–68. New York, NY: Springer.

117 DNV GL (2014). *A Review of Distributed Energy Resources*. Arlington, VA: DNV GL.

118 Honeth, N., Yiming, W., Etherden, N., and Nordström, L. (2011). Application of the IEC 61850-7-420 data model on a hybrid renewable energy system. In: *2011 IEEE Trondheim PowerTech, Trondheim, Norway*, 1–6.

119 Repo, S. et al. (2013). Active distribution network concept for distributed management of low voltage network. In: *IEEE PES ISGT Europe 2013, Lyngby, Denmark*, 1–5.

120 Jewell, W., Gomatom, P., Bam, L., and Kharel, R. (2004). *Evaluation of Distributed Electric Energy Storage and Generation*. Power Systems Engineering Research Center.

121 Wei, X., Zhang, J., and Yuan, Z. (2012). Research on the compensation capacity of the distributed energy storage in substation. In: *2012 China International Conference on Electricity Distribution, Shanghai, China*, 1–5.

122 Luo, X., Wang, J., Dooner, M., and Clarke, J. (2015). Overview of current development in electrical energy storage technologies and the application potential in power system operation. *Applied Energy* 137: 511–536. https://doi.org/10.1016/j.apenergy.2014.09.081.

123 U.S. Department of Energy (2002). Using distributed energy resources:a how-to guide for federal facility managers. *Cogeneration & Distributed Generation Journal* 17 (4): 37–68. https://doi.org/10.1080/10668680209508985.

124 El Oualkadi, A. and Zbitou, J. (2016). *Handbook of Research on Advanced Trends in Microwave and Communication Engineering*, 1ste, 716. IGI Global.

125 Information Resources Management Association (2016). *Renewable & Alternative Energy: Concepts, Methodologies, Tools, & Applications*, 1ste, 2006. IGI Global.

126 IEC 61850-7-420:2009 (2009). | IEC Webstore | LVDC, Webstore.iec.ch, https://webstore.iec.ch/publication/6019.

127 Ali, I. and Hussain, S. (2014). Control and management of distribution system with integrated DERs via IEC 61850 based communication. *Engineering Science and Technology* 20 (3): 956–964.

128 Apostolov, A. (2013). Integration of distributed energy resources. *Pacworld Protection Automation & Control World* 2013: 6.

129 Ahmed, M. and Kim, Y. (2014). Communication network architectures for smart-wind power farms. *Energies* 7 (6): 3900–3921. https://doi.org/10.3390/en7063900.

130 Zurawski, R. (2014). *Industrial Communication Technology Handbook*, 2nde, 1756. CRC Press.

131 Schwarz, K. (2012). NettedAutomation GmbH-The Net is the Automation. We bring people, systems and devices together, Nettedautomation.com, http://blog.nettedautomation.com/.

132 Pudjianto, D., Ramsay, C., and Strbac, G. (2007). Virtual power plant and system integration of distributed energy resources. *IET Renewable Power Generation* 1 (1): 10. https://doi.org/10.1049/iet-rpg:20060023.

133 El Bakari, K. and Kling, W. (2010). Virtual power plants: an answer to increasing distributed generation. In: *2010 IEEE PES Innovative Smart Grid Technologies Conference Europe (ISGT Europe), Gothenberg, Sweden*, 1–6.

134 Lundkvist, J. (2013). *Feasibility study of a Virtual Power Plant for Ludvika, Independent thesis Advanced level (professional degree)*. Uppsala University.

135 Sučić, S., Rohjans, S., and Mahnke, W. (2013). Semantic smart grid services: Enabling a standards-compliant Internet of energy platform with IEC 61850 and OPC UA. In: *Eurocon 2013, Zagreb, Croatia*, 1375–1382.

136 Pedersen, A., Hauksson, E., Andersen, P. et al. (2010). Facilitating a generic communication interface to distributed energy resources: mapping IEC 61850 to restful services. In: *2010 1st IEEE International Conference on Smart Grid Communications, Gaithersburg, MD, USA*, 61–66.

137 Dawidczak, H. and Englert, H. (2013). Integration of DER systems into the electrical power system with a generic IEC 61850 interface. In: *International ETG-Congress 2013; Symposium 1: Security in Critical Infrastructures Today, Berlin, Germany*, 1–5.

138 Wang, W., Xu, Y., and Khanna, M. (2011). A survey on the communication architectures in smart grid. *Computer Networks* 55 (15): 3604–3629. https://doi.org/10.1016/j.comnet.2011.07.010.

139 Rengaraju, P., Lung, C., and Srinivasan, A. (2012). Communication requirements and analysis of distribution networks using WiMAX technology for smart grids. In: *2012 8th International Wireless Communications & Mobile Computing Conference, Limassol, Cyprus*, 666–670.

140 Patel, A., Aparicio, J., Tas, N. et al. (2011). Assessing communications technology options for smart grid applications. In: *2011 IEEE International Conference on Smart Grid Communications, Brussels, Belgium*, 126–131.

141 Gungor, V. et al. (2011). Smart grid technologies: communication technologies and standards. *IEEE Transactions on Industrial Informatics* 7 (4): 529–539. https://doi.org/10.1109/tii.2011.2166794.

142 NETL (2008). *Modern Grid Strategy, Advanced Metering Infrastructure.* National Energy Technology Laboratory for the U.S. Department of Energy.

143 Vazquez, S., Lukic, S., Galvan, E. et al. (2010). Energy storage systems for transport and grid applications. *IEEE Transactions on Industrial Electronics* 57 (12): 3881–3895. https://doi.org/10.1109/tie.2010.2076414.

144 Emadi, A. (2005). *Handbook of Automotive Power Electronics and Motor Drives*, 1ste, 736. Boca Raton: CRC Press.

145 Yao, L., Yang, B., Cui, H. et al. (2016). Challenges and progresses of energy storage technology and its application in power systems. *Journal of Modern Power Systems and Clean Energy* 4 (4): 519–528. https://doi.org/10.1007/s40565-016-0248-x.

146 IEC (2003). *Communication Networks and Systems in Substations–Part 7-4: Basic Communication Structure for Substation and Feeder Equipment – Compatible Logical Node Classes and Data Classes.* IEC.

147 Ustun, T., Ozansoy, C., and Zayegh, A. (2011). Distributed Energy Resources object modeling with IEC 61850–7–420. In: *AUPEC 2011, Brisbane, QLD, Australia*, 1–6.

148 Ustun, T., Ozansoy, C., and Zayegh, A. (2013). Implementing vehicle-to-grid technology with IEC 61850-7-420. *IEEE Transactions on Smart Grid* 4 (2): 1180–1187. https://doi.org/10.1109/tsg.2012.2227515.

149 IEC (2008). *Communication Networks and Systems for Power Utility Automation for Distributed Energy Resources (DER).* International Electrotechnical Commission.

150 Bayram, I. and Ustun, T. (2017). A survey on behind the meter energy management systems in smart grid. *Renewable and Sustainable Energy Reviews* 72: 1208–1232. https://doi.org/10.1016/j.rser.2016.10.034.

151 Sauter, T. and Lobashov, M. (2011). End-to-end communication architecture for smart grids. *IEEE Transactions on Industrial Electronics* 58 (4): 1218–1228. https://doi.org/10.1109/tie.2010.2070771.

152 Gungor, V. et al. (2013). A survey on smart grid potential applications and communication requirements. *IEEE Transactions on Industrial Informatics* 9 (1): 28–42. https://doi.org/10.1109/tii.2012.2218253.

153 Parikh, P., Sidhu, T., and Shami, A. (2013). A comprehensive investigation of wireless LAN for IEC 61850-based smart distribution substation applications.

IEEE Transactions on Industrial Informatics 9 (3): 1466–1476. https://doi.org/10.1109/tii.2012.2223225.

154 Guise, L. (2012). *First Set of Standards*. Belgium: CEN-CENELEC-ETSI Smart Grid Coordination Group.

155 IEC (2007). *Power Systems Management and Associated Information Exchange : Data and Communications Security : Part 6: Security for IEC 61850*. Geneva: IEC.

156 Falk, H. (2008). Securing IEC 61850. In: *2008 IEEE Power and Energy Society General Meeting - Conversion & Delivery of Electrical Energy in the 21st Century, Pittsburgh, PA, USA*, 1–3.

157 Wang, Y. et al. (2011). Analysis of smart grid security standards. In: *2011 IEEE International Conference on Computer Science & Automation Engineering, Shanghai, China*, 697–701.

158 Fuloria, S., Anderson, R., Mcgrath, K. et al. (2010). The protection of substation communications. In: *SCADA Security Scientific Symposium*, 1, 13. Miami, USA.

159 Pathan, A. (2015). *Securiting Cyner-Physical System*, 426. Boca Raton, FL, USA: CRC Press.

160 Khosrow-Pour, D. (2017). *Encyclopedia of Information Science and Technology*, 4the, 8104. IGI Global.

161 Fan, Z. et al. (2013). Smart grid communications: overview of research challenges, solutions, and standardization activities. *IEEE Communication Surveys and Tutorials* 15 (1): 21–38. https://doi.org/10.1109/surv.2011.122211.00021.

162 Kalogridis, G., Sooriyabandara, M., Fan, Z., and Mustafa, M. (2014). Toward unified security and privacy protection for smart meter networks. *IEEE Systems Journal* 8 (2): 641–654. https://doi.org/10.1109/jsyst.2013.2260940.

163 Federal Communications Commission (2012). *Cyber Security Planning Guide*. FCC.

164 Sheng, Z.-Y., Qu, H.-P., Wang, C., and Zhou, X.-M. (2013). Research on security issues and solutions of IEC 61850 communication protocol architecture. *Journal of Theoretical and Applied Information Technology* 50 (2): 366–372.

165 McMillan, R. (2010). Siemens: Stuxnet worm hit industrial systems. Computerworld, http://www.computerworld.com/s/article/print/9185419/Siemens_Stuxnet_worm_hit_industrial_systems.

166 Silveira, M.G. and Franco, P.H. (2019). IEC 61850 network cybersecurity: mitigating GOOSE message vulnerabilities. In: *Proceedings of 6th Annual PAC World Americas Conference*, 1–9.

167 Hoyos, J., Dehus, M., and Brown, T.X. (2012). Exploiting the GOOSE protocol: A practical attack on cyber-infrastructure. In: *2012 IEEE Globecom Workshops, CA*, 1508–1513.

168 Senecal, L. (2009). Understanding, preventing, and defending against layer 2 attacks, Cisco Expo, http://www.cisco.com/web/ME/exposaudi2009/assets/docs/layer2_attacks_and_mitigation_t.pdf.

169 Yang, H.S., Kim, S.S., and Jang, H.S. (2012). Optimized security algorithm for IEC 61850 based power utility system. *Journal of Electrical Engineering and Technology* 7 (3): 443–450.

170 Ustun, T.S. and Hussain, S.M.S. (2019). A review of cybersecurity issues in smartgrid communication networks. In: *2019 International Conference on Power Electronics, Control and Automation, New Delhi, India*, 1–6.

171 Appendix E: Summary of NERC Cyber Security Standards.National Research Council (2012). *Terrorism and the Electric Power Delivery System*. Washington, DC: The National Academies Press https://doi.org/10.17226/12050.

172 Stouffer, K., Falco, J. and Scarfone, K. (2008). NIST SP 800-82: Guide to Industrial Control Systems (ICS) Security. Technical report. National Institute of Standards &Technology.

173 Farooq, S.M., Hussain, S.M.S., and Ustun, T.S. (2019). Performance evaluation and analysis of IEC 62351-6 probabilistic signature scheme for securing GOOSE messages. *IEEE Access* 7: 32343–32351.

174 Yunsuf (2008). *CCIE Profesional Development Series Network Security Technologies & Solutions*, 221. Cisco Press.

175 IEC standard 62351 (2007). Power systems management & associated information exchange, data & communication security.

176 IEC standard 61850.5 (2008). Communication networks & systems in substation, Specific communication service mapping.

177 Almas, M.S., Vanfretti, L., Singh, R.S., and Jonsdottir, G.M. (2018). Vulnerability of synchrophasor-based WAMPAC applications' to time synchronization spoofing. *IEEE Transactions on Smart Grid* 9 (5): 4601–4612.

178 Martin, K. (2011). Synchrophasor standards development - IEEE C37.118 and IEC 61850. In: *2011 44th Hawaii International Conference on System Sciences*, 1–8.

179 Harrington, H. and Anderson, L. (1998). *Reliability Simplified: Going Beyond Quality to Keep Customer for Life*, 384. New York: McGraw-Hill.

180 Yunus, B., Musa, A., Ong, H. et al. (2008). Reliability and availability study on substation automation system based on IEC 61850. In: *2008 IEEE 2nd International Power and Energy Conference, Johor Bahru, Malaysia*, 148–152.

181 Franco, P., Rocha, G., and Dolezilek, D. (2015). Case study: increasing reliability, dependability, and security of digital signals via redundancy and supervision. In: *PAC World Africa Conference, Johannesburg, South Africa*.

182 IEC (2009). *IEC 61850 Part 7-420 DER Logical Nodes - Communication Networks and Systems for Power Utility Automation for Distributed Energy Resources (DER)*. IEC.

11

Electric Vehicles in Smart Cities

Sahand G. Liasi and Mohammad T. Bina

K. N. Toosi University of Technology, Electrical Engineering Department, Tehran, Iran

11.1 Introduction

By the recent developments in Internet of things (IoT), the concept of smart cities and smart grids has become more widespread. IoT enables various devices to interact with each other and share their information. Processing this data in computing units, it is possible to add a portion of smart to the system. Electric vehicles (EVs) are emerging vehicles that are absorbing more and more consumers all over the world. Since EVs, which are supplied by electrical energy, can have great impact on the power system, it must be considered carefully in smart grids. EVs can exchange a great amount of information between themselves via IoT. Due to myriad number of EVs in cities, this issue must be considered seriously in smart cities. In this chapter, emphasizing the importance of EVs in smart cities, we provide information about various types of EVs based on their level of autonomous. Then, IoT technology and its relation with EVs will be explained. Since writing about IoT and connected vehicles without considering cyberattacks seems to be meaningless, we bring a great discussion to explain various types of cyberattacks that EVs face with. Finally, we focus on high penetration of EVs in cities and their impact on the power system. In this regard, we explain the influences on electricity demand and their role in demand response programs.

11.2 Autonomous Vehicle

Technologies related to autonomous vehicle (AV) are developing quickly. Recently, connected and autonomous vehicles (CAVs) and especially connected and autonomous electric vehicles (CAEVs) have attracted attentions in automotive industries.

Cyberphysical Smart Cities Infrastructures: Optimal Operation and Intelligent Decision Making, First Edition. Edited by M. Hadi Amini and Miadreza Shafie-khah.

Different manufacturers are investing in CAVs and developing various methods to make major changes in the automotive industry. In addition to major players, which mainly include traditional car manufacturers such as BMW and Mercedes-Benz, disruptive players such as Uber, which have no remarkable background in the automotive industry, are investing in this field.

Already, some traditional manufacturers such as BMW offer semiautonomous vehicles. These cars provide limited auto-drive options like self-parking or some other limited self-driving features. In the next few years, it is expected that CAEVs develop in ways that will have a substantial influence on transportation infrastructure and automotive industries. Highly reliable autonomous cars also may change vehicle's ownership issues in the future.

Vehicles also can be connected to each other so that they can exchange information between themselves. In this case, we call this kind of Internet of vehicles (IoV). This information exchange is not necessarily limited to data exchange between vehicles, but also they can exchange information with their surroundings via different communication media. To be more precise and to discriminate CAVs from other vehicles, for CAVs, we use IoCAV that is referred to as Internet of connected and autonomous vehicle.

There are various ongoing researches on CAEV due to their potential benefits. However, there are some challenges in developing this technology. For example, in smart cities with intelligent charging stations, which require reservation for charging EVs, CAEVs require automated reservation feature, enabling them to reserve a charging stand for themselves. Moreover, they need automated charging feature because in a smart city, it is expected that CAEV is able to reserve a stand in a charging station, go to the charging station, and charge its battery automatically without the presence of any people in the station. In other words, CAEVs are expected to be able to charge their batteries without any help of human or with the least aid by mankind. In this regard, CAEVs need to exploit their sensors, Global Positioning System (GPS), and inherent artificial intelligence to perform all the actions required for charging themselves. Moreover, wireless charging pads, which are also known as wireless charging station (WCS), are expected to be capable to provide automatic charging using wireless power transform (WPT) feature.

CAEV can be considered as a complicated autonomous or semiautonomous system, which consists of several recently emerged technologies to present all of their benefits in a package. Thus, CAEV can be introduced as a combination of connected vehicles (CVs), AVs, and EVs. To figure out the definition of each term precisely, here, we define each of these terms separately. A CV is defined as a vehicle that is able to communicate with its nearby objects. These nearby objects can be other cars or even any other type of objects that are able to connect to a CV using IoT. It is worth mentioning that these objects, which are equipped with IoT technology, can be smart or autonomous objects or any other objects. An AV can be defined as a vehicle, which is able to be driven without requiring human assist. However, in some contexts, vehicles that have the ability of semiautomated driving also are introduced as AVs. Finally, EV, as a general definition, can be defined as a vehicle that uses electrical energy as the main electromotive power. CAEVs can gather the received information from the environment and other EVs, process this data, and use them for improving the performance of themselves.

Generally, the components of a CAEV can be divided into five main categories [1]:

1. **Perception system**: This system senses the nearby objects to recognize the surrounding objects and provide a 3D map from the environment. For this sake, various sensors may be exploited such as radar, light detection and ranging (LiDAR), cameras, and so forth.
2. **Location system**: This system provides location information for the CAEV. For this aim, global navigation satellite system (GLONASS) is employed. Up to end of 2020, GPS, GLONASS, and BeiDou Navigation Satellite System (BDS) were operating as active GNSSs.
3. **Driving system**: This system is responsible for decision making during driving under various situations. This system decides how to avoid car accidents, changes lane for overtaking, etc.
4. **Communication system**: This system provides the communication infrastructure for connecting CAEV to its environment and other vehicles. The connection between the CAEV to other vehicles is called vehicle to vehicle (V2V) connection, while connection between the CAEV to other objects is referred as V2X.
5. **Energy system**: This system is composed of charging converters, battery, and energizing system in the vehicle.

11.2.1 Classification of Vehicle Automation

Society of Automotive Engineers (SAE) published J3016 standard, which classifies automated vehicles and introduces various autonomy levels. These autonomy levels include five levels, which defines no automation at level 0 to fully automated operation at level 5. To provide an example, consider the conventional vehicles that we use every day. This autonomy levels can be applied for them. The cheaper cars (and old cars), which do not provide any digital control assistant systems, can be categorized as level 0 autonomy. The newer ones, which are a bit more expensive, may provide some facilitating systems such as cruise control system. This results in a higher autonomy level. If, instead of simple cruise control system, adaptive cruise control system is provided, the level of autonomy increases. To be more precise about these levels, let us introduce these levels based on SAE definition [2]:

- **Level 0 – no automation**: The driver controls everything. No intelligent artificial control system is developed in the vehicle.
- **Level I – driver assistance**: At this level, some aids come from driving assistant, which can be speed control or acceleration control, for instances. Vehicles that merely enjoy adaptive cruise control system are categorized in this division.
- **Level II – partial automation**: More assists come from the driving assistance system. The assistant can control both acceleration of the car and steering control. However, the driver must monitor the road and the surrounding by himself. While the assistant is responsible for more tasks at this level, the driver still takes the main responsibility.
- **Level III – conditional automation**: The driving assistance can take responsibility of almost everything at this level. However, still it may request for driver's interaction by demand. While the driver can permit the automated control system to drive, they must be aware of critical situations, which need their intervention.
- **Level IV – high automation**: The control system can perform like level III automation system, but at this level, the system can continue its performance even if the driver does not response properly to the system requests. This can be assumed as low level of unsupervised driving system.
- **Level V – full automation**: The artificial intelligence of the automated control system takes responsibility for everything. It is able to monitor the road and environment and control the car under various conditions. Even it is capable to take responsibility under critical condition without any assistance by human driver.

11.2.2 Advantages of CAEVs

CAEVs provide various advantages in different aspects. Some of the most important privileges of this kind of vehicles are listed below:

- **Safety enhancement**: Using intelligent control system, which can autonomously drive the vehicle or help the driver (based on the automation level of the CAEV), it will attenuate the number of car accidents that are caused by driver error. It is estimated that at least half of the car accidents are caused by human error. Therefore, it is expected to reduce the number of car accident by a factor of 1/2.
- **Driving facilitation**: Today, using conventional cars, people have to learn how to drive and pass some standard examinations to receive a driving license. This is because of two main facts: first, driving conventional cars requires fundamental knowledge of working with cars (for example, how the steering function works or how changing manual gear works), and second, people have to learn how to

react as a driver to avoid accidents. However, in CAEVs, thanks to intelligent control service, even people with the least or no knowledge about driving will be able to run a car.

- **No/less need to parking space**: Nowadays, when people drive their cars to their destination, they look for a nearby parking lot or parking space to park their cars, but when they use CAEVs, they can get out of their vehicle and leave the car alone to find a parking space, maybe further than their location. Even it is possible to send the CAEV to home or any other defined location.
- **Taking rest in car**: While drivers must pay their full attention to driving when they are driving conventional cars, drivers of CAEVs can take rest when they drive CAEVs with high automation level.
- **Vehicle sharing**: Since CAEVs can be automatically go to different locations, it is possible to request for a shared vehicle easily. Therefore, there will be less need to individual cars, and car sharing will increase [3].
- **Decreasing traffic jam**: Using central control system, which CAEVs are connected to, the selected routes for vehicles are optimized. Also, employing various planning approaches, it is possible to avoid heavy traffic jams.
- **Reduction in air pollution**: Since CAEVs' batteries are charged by electrical energy, it can be stated that this kind of vehicles consume electrical energy instead of fossil fuels, which is used by conventional internal combustion vehicles. Since electricity can be produced by some other resources rather than only fossil fuels, air pollution reduction is expected. Even more, if the required electricity is generated based on fossil fuel power plants, since these power plants are located out of cities, at least we would have less urban air pollution.
- **Cost efficiency**: Electricity price is much lower than fuel price. Moreover, in many countries, governments have provided free electricity to encourage people to use EVs. Therefore, there will be a significant cost reduction in fuel cost of vehicle.

11.3 IoT Technology and CAEV

IoT can be used for various purposes and applications. The application for CAEVs is one of the most interesting ones. As explained above, IoT can be exploited for CAEVs to improve safety issues and automated driving. Moreover, utilizing IoT in cars (especially EVs) can aid further optimizations of smart city in various aspects. We will explain some of them in Section 11.4. As explained above, IoT provide data exchange not only between cars but also between other objects and smart city's infrastructure. To make the application of this data exchange and its effects on improving driving performance more clear, consider this example: when a CAEV (even at highest automation level) is running without

Table 11.1 Automation levels for vehicles introduced by SAE.

SAE level	Name	Execution of steering	Driving monitoring	Fallback performance
0	No automation			
I	Driver assistance			
II	Partial automation			
III	Conditional automation			
IV	High automation			
V	Full automation			

information exchange via IoT, it can sense different parameters such as vehicle's speed, ambient temperature, GPS data, etc. Also, it can use visual cameras to achieve some information about the surrounding, for example, recognizing road or road signs. However, this information is limited to the information sensed by the CAEV sensor and cannot obtain information about further distances. For example, it cannot achieve information about the traffic in its way to the destination. Using IoT, the CAEV, in addition to aforementioned data, can obtain more information from longer distance and can help it to optimize its way, driving method, and so forth. Moreover, IoT-enabled driving permits the car to have a better and more accurate information about the surrounding environment and perform appropriate reaction automatically. Recently, many governmental and nongovernmental agencies started and continued various research to develop IoV-based technologies. Table 11.1 illustrates a general overview of IoT and AVs. As we explained, and depicted in Table 11.1, the vehicle collects raw information from its environment by the sensors it is equipped with. This information will be analyzed and processed in computation unit of the vehicle. In addition, using data exchange over IoT communication, further information is inserted to the computation unit. Thus, analyzing both sensed data and IoT-based data, the control unit can achieve better performance and provide enhanced reactions against upcoming events. According to Table 11.1, external data sources include any IoT-equipped devices such as traffic lights, smartphones, vehicles, and so forth. Moreover, data exchange can be carried out over different communication technologies: 3G, 4G, Zigbee, bluetooth (BT), etc.

It is also worth mentioning that data exchange via IoT provides an alternative for vehicle sensors. This is because some data, which can be sensed by the vehicle itself, can be also be received by other sources. For example, environment temperature can be sensed by smartphones and then be sent to vehicles through the communication channel. Otherwise, this measurement can be performed by city center weather office and then be sent to the IoT-equipped units through the same communication channel or maybe another channel. At this condition, vehicle does not need environment temperature measurement unit anymore. This means that one sensor can be eliminated, which results in cost reduction. This example can be generalized to other sensors. Hence, more sensors can be omitted and the customer can enjoy reduced vehicle price. It must be mentioned that all sensors cannot be eliminated because some sensors are vital. As an instance, visual camera sensors cannot be omitted for fully automated vehicles because they are dependent to their camera for self-driving feature. Although most of the information, which is provided by the cameras, can be provided via data exchange using IoT, since self-driving feature is very sensitive and critical task, it cannot be merely executed over external data. This is because of two major rationales: first, external sources are not always available, which means self-driving feature cannot be always executable, and second, external information are exposed to hackers. Therefore, based on these reasons, CAEV cannot be merely dependent on external resources. However, external resources that provide such this information can be used as an alternative resource to improve sensed data or to be used under some conditions that vehicle's sensor are not working accurately or properly. In Section 11.6, when we will explain cybersecurity, more information will be provided about data exposure, hacked data, and possible dangers.

IoCAV is composed of three main parts: IoT integration unit, IoT platforms, and communication system. Diagram of the IoCAV platform is depicted in Figure 11.1. IoT integration unit is the main part of IoV, which enables vehicle to use the data exchange possibility over IoT communication system for self-driving and other features. In other words, this part integrates the whole IoT system of the vehicle. IoT platform includes IoT-enabled CAEV services like autonomous charging feature and fully automated taxi. Finally, the communication system is a network layer, which permits information exchange between different IoT-equipped devices utilizing communication protocols like Wi-Fi, 4G, long term evolution (LTE), and so on.

11.4 CAEV Applications and Services

Using the advanced features of CAEVs, it is possible to employ them for different aspects. In Sections 11.4.1 and 11.4.2, we will highlight two of the most important applications that CAEVs can be used in charging management system and taxi service. After providing a description on importance and general overview on these applications, we will allocate two separated sections (Section 11.5), in the latter parts of the chapter, to discuss the challenges in these applications.

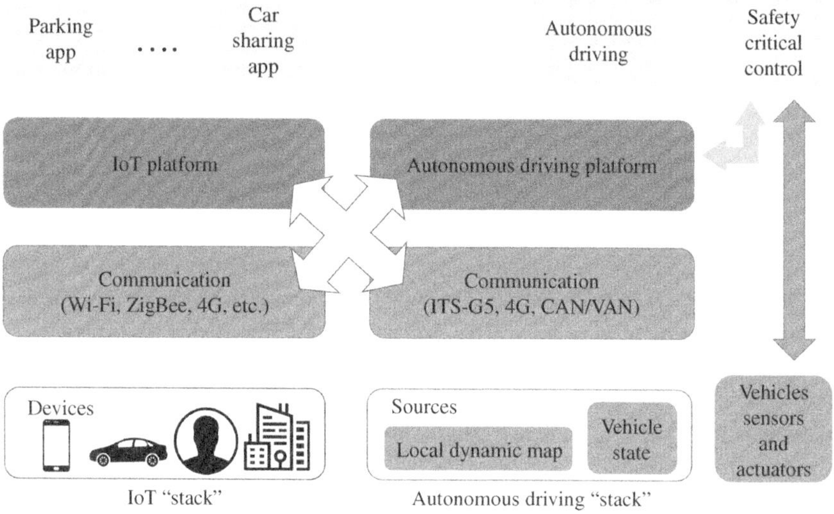

Figure 11.1 An overview of IoT and automation stacks for IoCAV platform.

11.4.1 CAEV Charging Management System

The process of charging EVs includes different actors: the EV owner, charging station operator, utility operator, etc. Replacing EVs with CAEVs, the role of EV owners will be omitted, and instead, artificial intelligence and central control center will be added as additional actors. This added actors work through IoT platform. CAEV charging management system is founded on the SecCharge platform. This platform has three main parts: the server, operator of WCS, and energy service provider. The simplified diagram of charging management system is illustrated in Figure 11.2. The energy service providers generate and distribute energy to both public and private customers autonomously. Public customers are consumers such as homes, and examples of these are charging stations, subway system, and universities. Operator of wireless charging station (OWCS) is not necessarily the operator of only one charging station, and it may operate several charging stations at a time. A specific EV charging network may have multiple operators of WCS. The SecCharge CAEV system is the main part of the architecture of the system, which plays the main role as the smart management unit for charging EVs. Supplying influential coordination with different OWCSs in EV charging system, it is designed to simplify charging services. Moreover, SecCharge server may help in finding wireless charger locations and managing CAEV data. This server is also responsible for the efficient interaction between OWCSs. In summary, it can be stated that SecCharge is an integrated server, which can manage various tasks related to EV's charging. At charging scheduling level, it can provide information about the empty and occupied charging spaces and also reserve charging spaces for CAEVs. Completing this task, it can assist charging automation procedure. Moreover, this

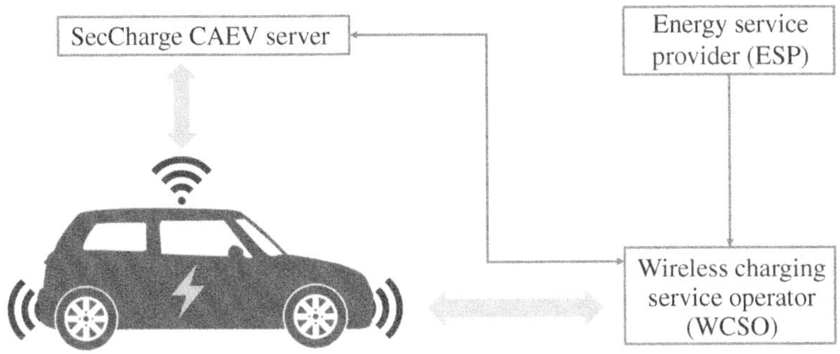

Figure 11.2 High-level diagram of CAEV charging management system based on the SecCharge platform.

prevents long queues at charging stations and traffic congestion. To be able to fulfill this duty, the server requires to receive information from EVs to determine their SOC to estimate the possible charging time. For this sake, a threshold (for example, 30%) can be defined so that when the SOC of the EV's battery reaches the threshold, the signal for charging will be generated. This level indicates an initial need for charging. To define a more restrict threshold, we need a critical threshold, which indicates urgent requirement for battery charging. Needless to say, the critical threshold must be less than the initial threshold. For example, 15–20% can be selected for this second threshold. While reaching the initial threshold sends signal to estimate the required time for charging, reaching the critical SOC level sends a signal that forces reservation of a charging space as soon as possible [4, 5].

11.4.2 CAEV Taxi Service

Using CAEVs for taxi services is one of the most profitable applications of CAEVs. The main goal of this type of taxies is taking the client and riding them to their destination. For this aim, the client sends a request over his/her smartphone or any other type of IoT-enabled device for a taxi. The central control unit allocates the nearest CAEV to the client. Then, the allocated CAEV taxi must go to the place of the person who had requested for a taxi and take him/her up. Using self-driving feature, the client will be driven to his/her destination. The cost can also be paid online. While at the background, there are many complicated tasks that must be completed to finish this process, the whole procedure seems very simple to the client, which is a great privilege because people prefer simple and fast service with a simple user interface.

In addition, spreading CAEV taxies in the cities, usual traffic jams and the need for parking space in the city will be attenuated. When riding from origin to destination is very cheap and easy using CAEV taxi, more people prefer to use this kind of taxies instead of their own car. Therefore, less private cars will run into streets,

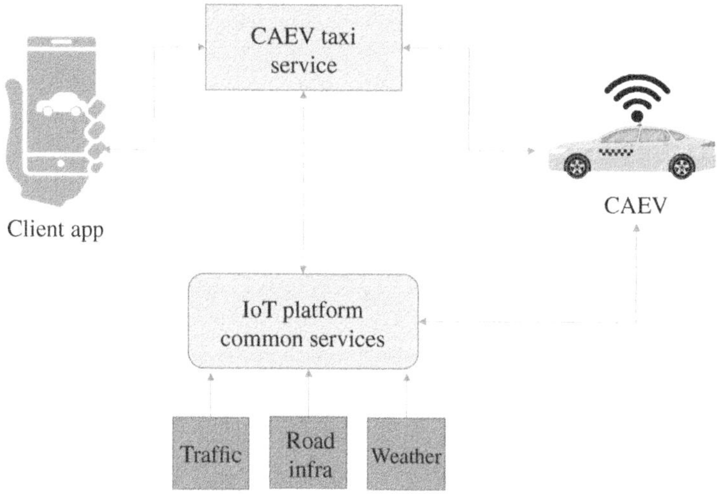

Figure 11.3 Requesting a CAEV taxi service from smartphone.

which leads to traffic reduction. Having less cars on the streets, less parking lots are needed.

Figure 11.3 illustrates a simplified diagram of CAEV taxi platform. As shown in Figure 11.3, in addition to the usual data that CAEVs require for automated performance, which is described in Sections 11.3 and 11.4, some extra information about the clients is transmitted through IoT network. This information is first analyzed and processed in central control center; then, the information will be sent to CAEV taxi. The sent information to CAEV taxi must at least include pickup and destination locations. The appropriate path from the pickup location to the destination can be computed by central control system and be sent to the CAEV taxi along with other information. However, it is possible for the CAEV taxi to find a proper route using its internal processing unit only by obtaining the initial locations. To enhance the performance of CAEV taxi, additional information, such as roads traffic data, geographical data, and some other basic information, may be sent to CAEV taxi from central control unit.

11.5 Cybersecurity Issues of Internet of EVs

The IoV, which is sometimes called connected vehicles, is an extended network that includes different devices such as vehicles, infrastructures, sensors, and so forth. These devices can send/receive data to/from each other to share their information about the surrounding environment. This information exchange helps vehicles to improve their driving performance by preventing undesired accidents and augment the safety level. Moreover, from macro perspective, the efficiency of the whole transportation system increases. IoV communication can be carried out via various means such as cellular technologies.

Since the energy stored in EVs' batteries is limited, various strategies are developing to deal with this problem. These approaches include both providing some ways to extend lifetime of batteries and increasing batteries' capacity and proposing methods to charge batteries more often. While improving battery capacity and their lifetime is mainly considered as a part of metalogics, we do not discuss this part in this chapter. However, using alternative solutions can be helpful. One of the approaches, as mentioned earlier, is providing more charging stations or providing other energy supplies for EVs. IoV can be used for this purpose. Employing the IoV, an energy map can be obtained to offer novel services for optimizing energy consumption of connected electric vehicles (CEVs). For this sake, CEVs exchange their obtained and sensed information with others and the central server. This data includes but not limited to the following: identification number, location, and state of charge (SoC). The exchange information is exposed to hazard of attackers. Thus, protection protocols must be taken strictly into account. It is worth mentioning that these attacks are not limited to stealing information, but also false data injection can be considered as a cyberattack in this case because false information results in malperformance of the main server. This malperformance can be limited to underestimation/overestimation of energy or can be extended to more serious malfunctions.

11.5.1 UPCEV Architecture

Urban Platform for Connected Electric Vehicles (UPCEV) tries to integrate data gathered by IoT and IoV to enhance EV-related problems. These problems include battery energy management, autonomous charging planning, the shortage of charging stations, and charging time. EVs on the street provide the UPVEC server (through IoT communication channel) with instantaneous data about EV's identification number, location information, ambient temperature, ventilation system status, weight, SOC of battery, traveled distance, and so forth (see Figure 11.4). The server processes this data and combines them with the recorded data about infrastructure (information about street holes, speed bump, etc.) and the received data from satellites about road traffic and weather condition to build an energy map, which includes information related to EV power consumption. The created map can be employed to suggest new services to EVs. For example, it can offer routes, which require lowest energy to reach the destination [6].

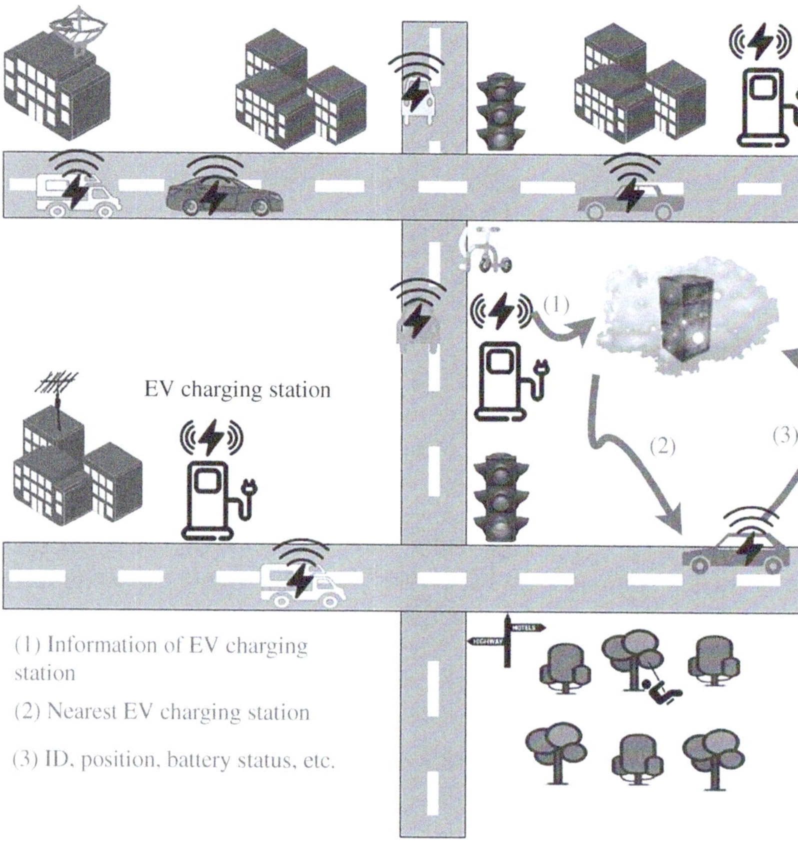

Figure 11.4 UPCEV architecture.

11.5.2 IoV Communication Technologies

The IoV give the vehicles the power to communicate with both interior and exterior ambient. Interior ambient consists of sensors of vehicles, while exterior ambient includes every other objects out of vehicle such as other vehicles, people, and infrastructure. Figure 11.5 illustrates the architecture of the IoV for EVs, which will be called IoEV from now.

To describe different communications precisely, in this subsection, we will introduce V2X communications [6]:

- **Vehicle to sensor (V2S)**: V2S aims to connect a vehicle to its embedded measurement devices. Sensors target to measure and obtain data about the condition of the vehicle and its surrounding ambient. Gathered information will be forwarded to the power control unit for additional connection between the vehicle

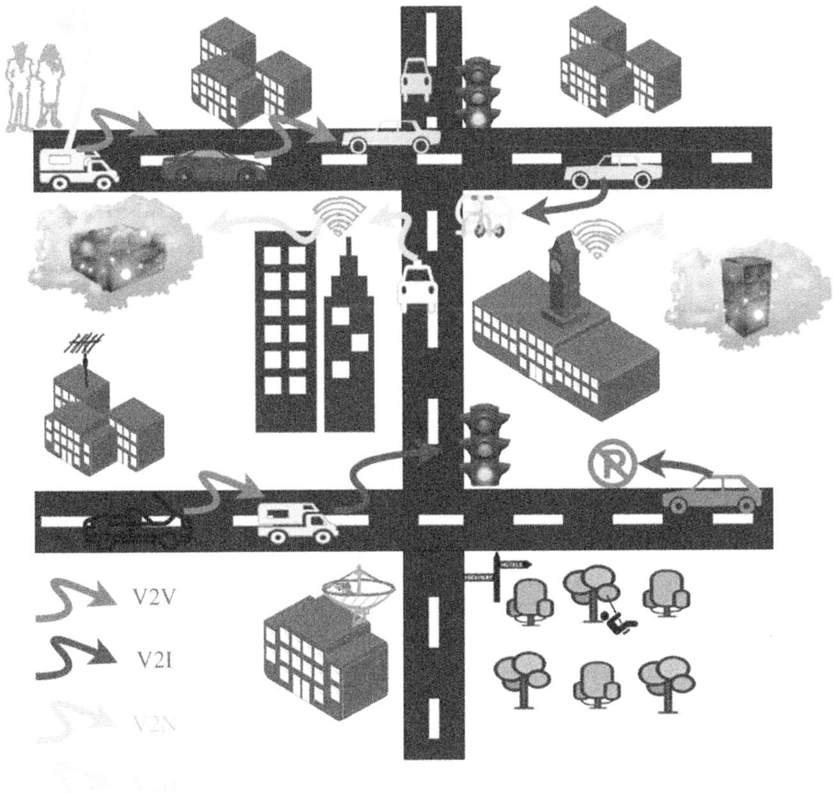

Figure 11.5 Architecture of IoEV.

and other objects. This type of communication can be established on both wired and wireless communication. In fact, measuring devices and the engine control unit can commune via wire-based systems, such as Local Interconnect Network (LIN), Controller Area Network (CAN), and FlexRay, or via wireless-based systems, such as Zigbee and Wi-Fi. It is also worth mentioning that due to the cost efficiency, wireless communication is mostly preferred to the wired one for V2S.

- **Vehicle to vehicle (V2V)**: This kind of communication permits vehicles to connect and share information with each other. Proximity Services, which are usually called ProSe, is almost the most common technology that certifies the V2V connection according to the IEEE 802.11p standard. V2V is a kind of device-to-device (D2D) communication, which permits two devices to share information with each other in an authorized cellular bandwidth.
- **Vehicle to road (V2R) or Vehicle to infrastructure (V2I)**: V2I/V2R allows vehicles to connect with the infrastructure like roadside units and road signals.

This helps vehicle to attain information about road condition or traffic from distributed sensors in the roads. Like V2V, V2R communications is mainly founded on IEEE 802.11p standard.

- **Vehicle to internet (V2I) or vehicle to network (V2N)**: This communication category delivers a straight communication among vehicles and the Internet. As expected, V2I is founded on hotspot and cellular technologies. The main goal of this type of communication is facilitating access to the Internet. The most recent cellular technology that can be employed for V2I purpose is 5G. Thanks to higher speed provided by this generation of technology standard for broadband cellular systems, the access to the Internet can be facilitated.
- **Vehicle to human (V2H) or vehicle to pedestrian (V2P)**: This type of information exchange allows data exchange between vehicles and humans. Through this type of communication, in addition to drivers, passengers and other pedestrians can interact with vehicles through their IoT-based devices.

11.5.3 IoEV Vulnerabilities

In the following, we will introduce security problems of IoEV communications and underline their influence on the UPCEV system. IoEV is a sophisticated infrastructure that includes heterogeneous nodes such as sensors, vehicles, and so on. Moreover, it consists of various kinds of communications. To have a more precise investigation about different kinds of communications, the attacks are divided into four categories, and each one is described in details.

11.5.3.1 V2S Attacks

Firstly, we scrutinize the security problems of V2S for both wired and wireless communications and emphasize their effects of UPCEV. Generally, communications based on wire are vulnerable to malware. This vulnerability permits hackers to control the CAN bus. As a general inference, it destructs vital services on the vehicle. Attacks to the network of wireless sensors may be carried out on the interior network of vehicle. An exception to these attacks is vehicle's sensor that utilizes EV battery as their energy supply. Actually, when sensors are connected to the main battery of the EV, they almost have no limitation related to energy supply. Although this statement is usually true, in some cases that SOC of battery is very low and there are no nearby charging stations, vulnerability to these attacks arises. A hostile node can gravely damage the vehicle performance by activating other attacks on the vehicle sensor network [6]:

- The **jamming attack** may be activated to stop proper performance of sensors' data exchange with the electronic control unit, and consequently, the information exchange between vehicle and the server will face serious problem. Moreover, the self-driving system cannot estimate vehicle conditions accurately.

- **False data injection attack**: This kind of attack, more than physical layer, concentrates on network layer to intercept data transmission and substitute the correct information sent from sensors with random or intentionally manipulated data to disrupt the appropriate performance of the vehicle. This false information can be false data about everything. For example, if the attacker injects false SOC data about the vehicle's battery, say, illustrating the battery level lower than its actual value, the automatic driving system tries to consider finding a charging station to charge vehicles battery. Therefore, a simple change in only SOC of vehicle can change the ultimate goal of vehicle (its destination). Similar scenario can happen if the hacker injects false information about other parts. Even if the vehicle is not using auto-driving feature while misinformation is being shown to the driver, it convinces the human driver to make a wrong decision merely because of false data. At a worse condition, if false data coming from hackers were about ongoing traffic, the driver may change his predefined route to the destination to avoid unwanted traffic. By a maliciously biased information, attackers may try to conduct the vehicle to their desired place to steal the vehicle. Even at the worst condition, this attacks may cost the driver's life.
- **GPS deception**: In this kind of cyberattacks, false GPS information is injected, which results in malfunction of vehicle in location-related tasks. Clearly, the route selected by control unit of the vehicle to drive it to the defined destination is highly dependent on the location service. While wrong information is fed to the control unit, the vehicle will be unable to determine an appropriate path to its destination. This also influences the energy consumption of the EV because the vehicle cannot find the optimal path. The problem will be worse at rural roads, where the available charging stations are less than those in the city.
- **Denial-of-service (DoS) attack**: While, in the previous type of cyberattacks, attacker was trying to mislead the vehicle by false information, at a DoS attack, attacker bombards the electronics control unit with a myriad number of pointless messages to disturb the performance of the control unit. This can prevent the process unit to be able to gather data and send them to UPCEV server.

11.5.3.2 V2V Attacks

In this subsection, we introduce V2V attacks that can happen for UPCEVs [6]:

- **Selfish attack**: This happens in the case where an EV refuses to cooperate with other EVs to share or relay the information because of its own benefits. For example, an EV may prefer to reduce its power consumption rate; therefore, it rejects any requests for information exchange. In another case, the EV maybe has direct access to the Internet and does not want to cooperate with nearby vehicles. If the number of these selfish vehicles increments, more selfish nodes appear at the network, and this may result in malfunction of the whole network.

- **Modification attack**: This kind of attacks is very similar to "false data injection attack," which introduced for V2S. In this type of attacks, the attacker modifies the content of information sent or received by vehicles. As an instance, an EV shares its information about its velocity, SOC, etc. with the server to be processed in central control unit and request to receive information about the best path to its destination. Then, the control unit will send a response to the vehicle to comply the request. Here, we have two signals: one from the vehicle to server and another from server to the vehicle. An attacker may intercept each of this signals to modify exchanged data. If the attacker changes the sending data from the EV, the control unit will compute the best route based on false information; therefore, its response to EV will not be correct, which misleads the EV. Moreover, if the data is sent to the server without any modification, but the attacker alters the correct response sent from the control unit to the EV, again, EV receives misleading information.
- **Sybil attack**: A malicious node may introduce itself with several identities. In this case, the server receives wrong information about the system, and its analysis outcome demonstrates a wrong demonstration of the whole system. For example, a malicious driver wants to have its own way not to be crowded and introduces its vehicle as tens of cars instead of one vehicle. In this case, the control system identifies that specific road with more traffic than what it is in the real world. Therefore, system marks the road with heavy traffic jam or in worse cases, marks the road as a congested road. Consequently, other vehicles, which are requesting the control unit to propose the best way to them, receive a response in which that specific road has been replaced by another alternative one. Thus, less number of vehicles will go through that street, making the road free for the malicious driver, while other have to spend more time to reach at their destination. Another example can be about charging stations. If a malicious vehicle with more than one identity goes to a charging station, the charging station will be considered a station with long queue; therefore, other drivers who are looking for a charging station decide not to go to that specific charging station. All in all, at this kind of attacks, the intelligent transportation manager, which is installed at central unit, is misled by the malicious vehicle, which consequently results in providing misinformation for the other vehicles.
- **False data injection attack**: A malicious vehicle may forward false data to its nearby vehicles to impact on estimation of the UPCEV server.
- **Eavesdropping**: An unauthorized node can eavesdrop to perceive critical data about other vehicles, which can result in privacy attacks. The critical data include but not limited to vehicles' location, history of travels, and identification.
- **Black hole attack**: This type of threat in very common in vehicular ad hoc networks (VANET). The target of the attacker is to put the sender into mirage,

which it is a property of to the shortest route to catch the transmitted data. Thus, the attacker may drop the message and reject to broadcast it that can lead to a refutation of service attack on the system when the system could not gather enough data to predict energy.

- **Gray hole attack**: This kind of attacks is an extended type of the previous one. Although the behavior of the attacker in both gray and black hole attacks is the same, the gray hole attack is more dangerous because the attacker can drop messages selectively. As an instance, a malicious node may try to send all types of messages excluding traffic safety messages.
- **Wormhole attack**: The attacker wants to repeal legal links between vehicles. Generally, malicious bodies, which are sited in various geographical places, may conspire to send packets to uncertified area. Consequently, any transmitted information between two authorized vehicles will go through the attacker. Therefore, when an attacker obtains a message in a legible network point, it confines it and sends it to another attacker to revive this message in the network.
- **DoS attack**: The attacker tries to stop appropriate performance of the system. This disturbance can result in a partial disturbance complete shutdown. This attack does not permit EVs to communicate with each other.

11.5.3.3 V2I Attack

For connecting to a UPCEV server, vehicle requires obtaining an IPv6 address. The IPv6 addresses of vehicles are consistent of a network prefix, which is forwarded by roadside units (contemplated as a router in the system) and an interface recognizer, which is built in an extended unique identifier (EUI-64) format encoded in 64 bits. The EUI-64 interface recognizer is taken from the vehicle's media access control (MAC) address. Each roadside unit frequently sends router advertisements, which include the IPv6 prefix to all vehicles placed in its area [6]:

- **Replay attack**: The attacker repeats a router advertisement from the past to neighbors. Thus, victims cannot connect to the legal roadside unit when the internet protocol (IP) next hop address is not correct. For UPCEV, this attack does not allow the server to gather data from EVs to upgrade the map.
- **Router advertisement forgery**: A malicious node can fake a router advertisement message with a void next hop address not permitting its neighbor nodes to have access to the Internet. Consequently, they will not have access to server. Therefore, vehicles cannot connect to neither the roadside unit nor the server.
- **Privacy attacks**: Even if we assume that the sensed data have been sent to the server without any changes and also the same desired transmission happens for the server-to-vehicle message, there is still the chance for a malicious node to create a map based in the data transmission over a relatively long period.

The attacker can track a car by its temporary address. Not only this can jeopardize vehicle's owner privacy, but also this information can be exploited to plan a physical attack.
- **RSU spoofing**: This kind of attacks includes spoofing the IP address of the RSU and then forwarding a void prefix. It merely influences the mobility when the car configures with a void temporary address.

11.5.3.4 V2N Attacks

V2N communication is founded on cellular technology (say, 4G as an instance), Wi-Fi access points, and sometimes satellite communication. In this subsection, a general overview of security topics of V2N for UPCEVs will be provided.

An attacker can carry out a DoS attack by uploading a myriad number of successive messages to the access point, like the bombardment attack explained earlier. This huge number of messages can fill the bandwidth of the Wi-Fi access point. Attacks like man in the middle can be prevented using Passpoint/Hotspot 2.0. MOBIKE is the most common security protocol utilized by Mobile Femtocell. It permits key exchange among the Mobile Femtocell and the core network and IP multi-homing without reconstructing all security associations and IPsec tunnels.

11.6 IoT-Based EV State Estimation and Control Under Cyberattacks

Transportation system is developing toward high-automation systems. Nowadays, trains in some subway systems, like ones in Japan and India, are running without driver. However, such these systems are very limited. Moreover, driverless systems are only limited to train application, which requires relatively simple control system in comparison with taxies or urban buses because there are not many lanes to drive in, and the trains only have to be controlled in terms of speed and safety. This story is not as simple for taxies. Taxies have to drive in crowded city streets, find the passengers beside streets, take them, and ride them to their destination. Considering different routes from origin and destination makes the problem more challenging. Furthermore, for train, all of the trains in subway system of a city can work integrated and be guided with both dispersed and central control units, while for taxies, in addition to these units, there are many other vehicles that are driven by humans, which means predicting their behavior is more complicated. In this regard, artificial intelligent is considered as a well-applicable solution for this problem. Using artificial intelligence, it is possible to enhance the driving system in a way that all the aforementioned problems can be solved. Intelligent transportation system has become one of the most interesting topics at both academic and industrial levels. This will be more sensible when we remind that a great portion of the price we pay for taxies is for their drivers. Now, assume driverless taxies. There is no cost for driver. Clearly, the transportation cost will reduce significantly. On the other hand, the companies that are operating this system can run too many taxi with a limited number of employees and all tasks will be performed by computers, a huge benefit for the company.

Since here we emphasized the importance of CAEV role in public transportation, we must address the challenges as well. Since running CAEVs is directly related to the safety of people, governments take intelligent transportation seriously and regulate strict rules for them. Therefore, there should be many efforts on enhancing driving safety for such systems. In addition, ensuring the privacy of the users is another paradigm. To build a fully automated vehicle, which can be considered driverless, the sensing and communication units are obligatory. This is because EVs and central control units are located far away from each other. As depicted in Figure 11.6, the information obtained by EVs' sensors are sent through IoT communication network to the control center. Cyberattacks can occur in communication network when the information is being sent from the sensor to the control center or from control center to the CAEV. These attacks may aim to steal the information or to inject misleading information. Nevertheless, this can lead to sabotage by hackers. State estimation is carried out at the central control unit based on the data received from CAEVs. This information portrays a snapshot of the connected vehicles' operating condition at each time. Thus, it can be stated that using this state estimation, the physical system can be visualized. This estimation helps the control center to take required actions. Needless to say, state estimation performs a pivotal role in appropriate performance of the control system. In the following, we will explain an algorithm to monitor and control CAEVs via IoT under cyberattacks.

Fortunately, IoT technology can be embedded into a wide variety of devices, from small devices such as smart wristbands and smart watches to large ones like EVs. All of the sensed and measured data will be sent through IoT channel to central control unit. This concept is illustrated in Figure 11.6. This information on its way toward central control unit is exposed to hackers' hazard. Therefore, the information is vulnerable to cyberattacks, leading to security and privacy concerns.

Different algorithms are introduced to monitor and control EVs. Among them, approaches based on Kalman filter algorithm and Luenberger observer have gained attentions recently. State estimation of the system under cyberattack condition is also considered in the recent literature.

11.6.1 State Estimation Problem

State estimation of EVs is highly dependent to the data sensed and measured by EVs because all the computations are based on this information. In addition to

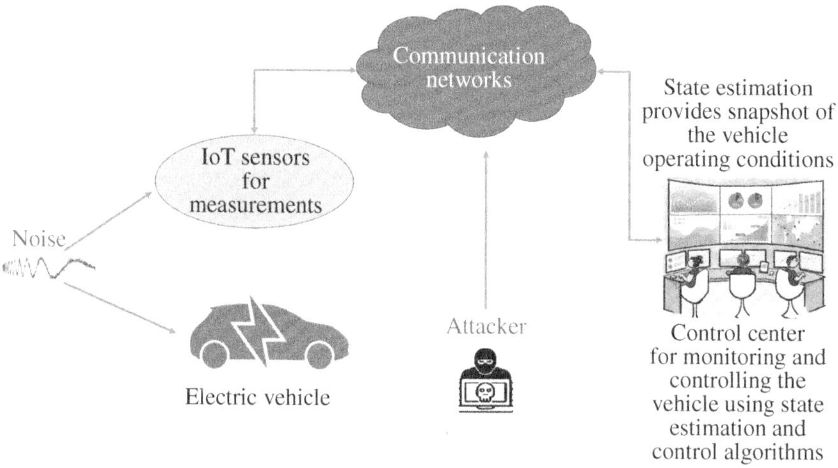

Figure 11.6 An overview of IoT-based EV and control center while an attacker is performing a cyber attack.

accuracy of the sensed data, which may affect the final performance of the whole system, cyberattacks are another effective factor. Measured data is transferred through IoT communication channel. This channel is vulnerable against hackers' attacks, which has caused many concerns about security and privacy aspects [7]. Thus, it can be inferred that attack recognition and protection against them are serious challenges to warrantee the resilient performance of the IoT-based system.

According to this issue, in Sections 11.6.2 and 11.6.3, we will discuss optimal state estimation while the IoT sensing data is under cyberattack. This discussion include how state estimation operation can withstand cyberattacks and which control scheme can be employed to adjust the system states. In this regard, the condition in which attackers inject misinformation to the network to force the control system to make mistake will be explained. For this aim, first, state-space framework is introduced, and then discussion is carried out.

11.6.2 Vehicle State Space and IoT Sensing Systems

With increasing number of vehicles in roads and streets, drivers usually face heavy traffic jams. This issue is more serious in large and populated cities. The heavy traffic jam results in drivers' anger, which increases the risk of misbehaving, and, consequently, car accidents. Vehicle automation systems assist drivers under such these conditions. As we explained in Section 11.2.1, the assistance can be limited to some basic tasks or be developed as full automation. Recent research studies have proved that these systems decrease the workload of the driver, which finally leads

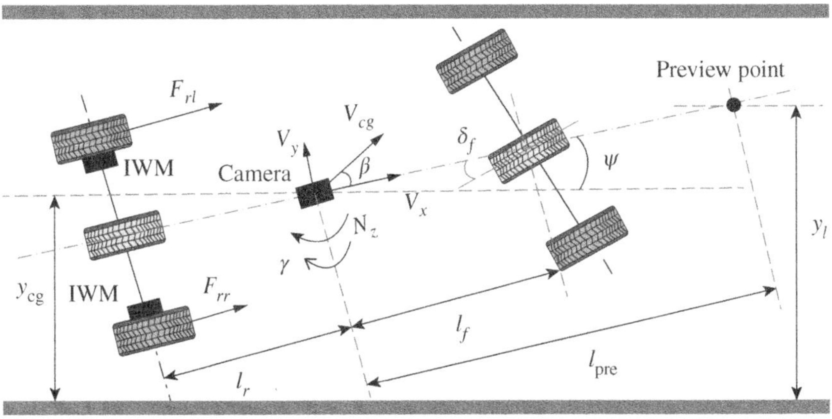

Figure 11.7 Combined vehicle and visual system.

to a significant reduction in road accidents. To have such a system, an accurate control scheme must be designed.

In this book, it is assumed that the model of vehicle is integrated with visual camera system. Figure 11.7 shows this integrated model. As shown in Figure 11.7, four main state variables to describe vehicle's behavior. It must be mentioned that a slip angle needs to be defined here because the actual direction of vehicle motion differs from the direction of rolling wheels. The best location for the camera is in front of the windshield because in addition to monitoring the objects in front of car, it permits the camera to see the shape of lines and identify the position. The angle of the front wheels is required for direction control of the vehicle. For designing a comprehensive control algorithm, the interaction between the vehicle's dynamics and visual monitoring system is required to be modeled in a state-space frame work. For providing various IoT-enabled services, which are explained in Section 11.4, the system operator deploys a set of IoT-enabled sensors.

11.6.3 State Estimation Under Cyberattack

In the following, an optimal state estimation and control algorithms will be introduced to monitor and control the IoT-based EV under cyberattack. We explain the vehicle dynamic with onboard visual system in state-space framework. We consider hackers' cyberattack in the IoT communication channel while the measured information is sending to the central control unit.

Vehicle's differential equations can be stated as [8]

$$\dot{\beta} = 2\frac{C_f}{mV_x}\left(\delta_f - \gamma\frac{l_f}{V_x} - \beta\right) - \frac{\gamma}{mV_x} + 2\frac{C_f}{mV_x}\left(\gamma\frac{l_r}{V_x} - \beta\right) \tag{11.1}$$

Table 11.2 Parameters of differential equations.

Parameter	Description	Parameter	Description
β	Body slip angle	V_x	Vehicle speed
C_f and C_r	Front and rear tires cornering stiffness	δ_f	Angle of the front wheel
m	Mass of vehicle	γ	Raw rate
l_f and l_r	Distance between the center of gravity and the front and rear axle	d_r	Track width
I	Inertia	N_z	Yaw moment
T_l and T_r	Rear left and right IWM torque	r	Radius of wheel

Source: Modified from Rana [8].

$$\dot{\gamma} = 2\frac{l_f C_f}{I}\left(\delta_f - \gamma\frac{l_f}{V_x} - \beta\right) - \frac{N_z}{I} + 2\frac{l_r C_r}{I}\left(\gamma\frac{l_r}{V_x} - \beta\right) \tag{11.2}$$

The torque produced by in-wheel motor (IWM) can be formulated as

$$T_l = F_{rl} r = \frac{mra_x}{2} + \frac{rN_z}{d_r} \tag{11.3}$$

$$T_r = F_{rr} r = \frac{mra_x}{2} - \frac{rN_z}{d_r} \tag{11.4}$$

where the parameters of these equations are given in Table 11.2.

Using above equations, discrete-time state-space framework can be attained as follows:

$$x_{k+1} = A_d x_k + B_d u_k + n_k \tag{11.5}$$

where $x = [\beta\ \gamma\ \psi\ y]'$, $A_d = e^{A_c T}$, $B_d = \int_0^T e^{A_c \tau} B_c d\tau$, $u = [\delta_f\ N_z]'$ and n_k represents the noise in the kth time step.

To monitor EVs, systems operators implement smart sensors based on IoT technology, which can be described as follows:

$$y_k = Cx_k + v_k \tag{11.6}$$

where y_k is the observation data at the kth time step, C is the measurement matrix, and v_k is the sensing noise at the kth time step. The sensor analyzes the raw data, locally. Then, the sensed data, which is the true data and can be expressed as $z_k = y_k - Cx_k^-$, will be transmitted through the IoT channel, where x_k^- represents previous state data. The manipulated data, injected by a hacker, can be stated as $\tilde{z}_k = T_k z_k + a_k$, where T_k is the attacker matrix and a_k represents channel noise.

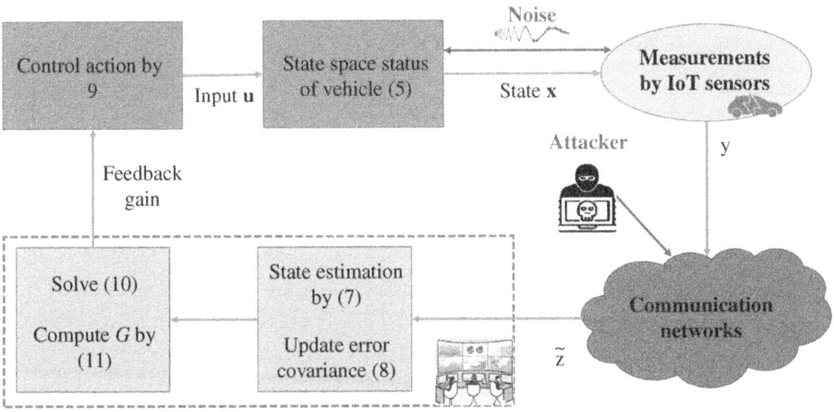

Figure 11.8 Overview of IoT communication systems and algorithms.

The injected data by attackers is targeted to mislead the central control unit. Therefore, the algorithm must be designed in a way to stabilize the system state. The general overview of the process is presented in Figure 11.8.

According to the mean squared error (MSE) concept, the optimized state estimation algorithm must be extracted to visualize the EV states. For a specific state-space framework in (11.5) and measurement in (11.6), the state estimation of the system can be established based on the following theorem:

$$\widetilde{x_k^-} = A_d\widetilde{x_{k-1}}, \quad \widetilde{x_k} = \widetilde{x_k^-} + K\widetilde{z_k} \tag{11.7}$$

The updated error can be calculated as follows:

$$\widetilde{P_k^-} = A_d\widetilde{P_{k-1}}A_d' + Q, \quad \widetilde{P_k} = \widetilde{P_k^-} + \overline{P}C'(\check{P} - T_k'\check{P} - \check{P}T_k)C\overline{P} \tag{11.8}$$

where $\check{P} = (C\overline{P}C')^{-1}$. To find gain factor, which will be sent to smart actuator, we can write the inequality equation as

$$\begin{bmatrix} -X & XA_d' + S'B_d' & X \\ X(XA_d' + S'B_d')' & -X & 0 \\ X & 0 & -\xi I \end{bmatrix} < 0 \tag{11.9}$$

where $\xi > 0$ is a programming variable that optimizes the estimator error covariance.

This is a linear matrix inequality in X and S. After solving (11.9), the optimal gain can be obtained according to the following equation:

$$G = SX^{-1} \tag{11.10}$$

11.7 Effect of EV Charging Behavior on Power System

Load profile shows electrical demand of loads and can be sketched in scale of day, week, month, or even year. Taking a closer look into load profile, clearly, it can be inferred that the load curve is not a flat line, but it has some ups and downs. This is because electricity demand is not constant in all time periods. For example, in summer, when temperature goes up, people start using ventilation systems for cooling their houses, workplace, etc. This means additional load demand in summers. However, lower electricity demand can be expected during spring due to more continent temperature. This demand variation is not only for long-term periods of times, but also load demand varies every day and every hour. Figure 11.9 illustrates a typical load profile of 24-hour time horizon. According to Figure 11.9, in some hours, less power demand exists, which is shown in gray. Such this time period is usually called valley, due to their valley-like shape. Also, there is some hours that the power grid experiences high power demand, which is called peak hours. Some literatures have defined an additional category, which is called critical peak. Critical peak introduces the peak hours that the load demand surpluses the available power generation capacity, endangering the reliability of power grid. The remaining hours, which are indicated by light gray in Figure 11.9, are called off-peak hours.

High penetration of EVs in smart cities will result in higher power demand for the electricity system because EVs must be charged by electrical energy. While EV owners plug their EVs into the grid without demand considerations, they would jeopardize the whole power system unconsciously. This is because they may try to charge their EVs during peak hours. Needless to say, connecting too many EVs

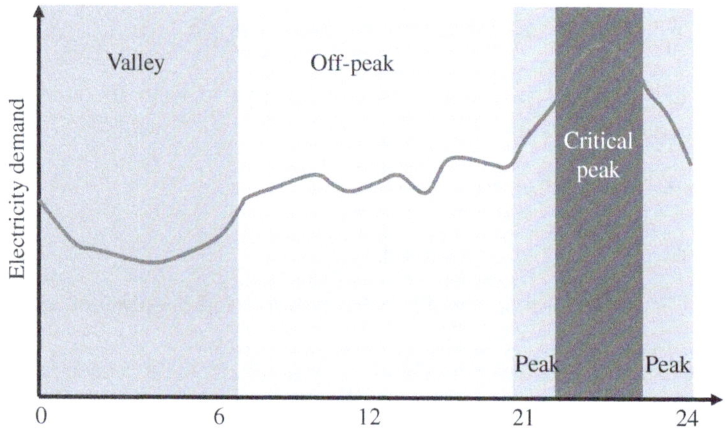

Figure 11.9 A typical daily electrical power demand.

at once will increase power demand hazardously. To avoid this undesired effect, many research has been conducted to propose approaches for preventing this phenomenon. On the other hand, if EV owners try connecting their EVs to the network during valley hours, they can help the power system operator to run the system more efficiently.

In addition, considering the available energy stored in EVs' batteries, it is possible to inject energy from EVs to the main grid during peak hours, which can help system operator to operate the system with higher reliability.

To investigate these influences, in Sections 11.7.1–11.7.3, we will explain the effect of demand response programs on EVs charging and electricity market operation.

11.7.1 Behavior of Electric Vehicle Parking Lots as Demand Response Aggregation Agents

Enhancing EV technologies offers different opportunities for improving power system reliability and performance. More than improving system efficiency and performance condition, fossil fuel consumption and air pollution will be decreased when EVs find their place in transportation system. In addition, because of advantages of demand response for obtaining reliable and efficient electricity markets, demand-side management programs are considered as a pivotal component on the sustainable development path. In this regard, it is important to find an efficient structure of demand response aggregation in electricity markets [9].

Different kinds of demand response programs have been defined to cope with peak demand problem. Considering the presence of EVs in smart cities, it is possible to take EVs' battery capacity into demand response programs. Due to the significant capacity that aggregation of these batteries provides, this capacity is very attractive to power systems operators and researchers. At the first glance, high penetration of EVs endangers power systems' reliability because they require remarkable power demand to fill their batteries. While there is no control on EVs, charging schedule, the owners of EVs can plug in their EVs whenever they want to charge their vehicles, which may result in new peak demand that is greater than current demands. This problem gets more serious when we understand that people prefer to charge their EVs during noon, when they have left their cars at parking lots, or during evening, just after reaching home. This is a serious challenge because these two periods of time are coincident with peak hours [10]. To attenuate this problem, smart charging schedule can be implemented to avoid peak demands. Moreover, when it is possible to inject power from EVs' batteries into the main grid, not only it is possible to prevent new peak demands, but also peak shaving is possible.

Therefore, integration of EVs into the power grid is vital. One of the best solutions considering the integration is influential operation of EV parking lots.

However, charge and discharge management would be a challenging paradigm from the parking lot owners' perspective. There are two main factors that cannot be determined by arithmetical approaches because of their probabilistic inherent: stochastic behavior of EV owners and uncertain electricity price at electricity market. While signing tariff-based contracts with the operator can reduce the risk, it constrains the income of this new player in the electricity market. If the parking lot owner takes the risk of unpredictable electricity price, it has the chance to take part in the market as a flexible player in demand response programs. Thus, it can be stated that when EV parking lots take part in demand response programs, both owners and market operators will receive higher benefit.

On the other hand, the EV owners prefer to keep their cost at the lowest level. Hence, the aggregators are third-party players, which participate in the market on behalf EV owners. Furthermore, distribution system operators try to manage EV charging scheme to avoid additional peak increase and line congestions. There are many optimization approaches that are proposed to maximize the benefit in demand response programs. Statistical model has been used extensively to model the stochastic behavior of EV owners. Some advanced methods have also been employed for this sake, including machine learning and neural network approaches.

Demand response programs attempt to force consumers to react to alterations of electricity price in various hours. Generally, these use incentives and penalties to encourage consumers to change their power usage algorithm. By participating in these programs, both consumers and the system operator can achieve their own benefits.

11.7.2 Parking Lots in Demand Response Programs

For modeling the behavior of parking lot operators, the energy exchange between the parking lot and the main grid must be computed. However, this cannot be carried out as easy as it seems. Various stochastic and scenarios must be taken into account to obtain an accurate calculation. To be more precise, patterns of EV owners' behaviors must be extracted to find about their departure and arrival times. Furthermore, various scenarios for electricity trading have to be considered. The operator or owner of the parking lot tries to increase their benefit from power exchange between the EVs and the power grid. However, in this way, they must consider EV's minimum requirements. For instance, it is not possible to inject the entire available energy of EVs to the network because drivers need this energy to drive their cars from the parking lot to their destination. Moreover, it must be considered that parking lots have two main constraints: time limitation and energy limitation. This is because EVs are only parked at parking lots for a limited time period, and according to the number of EVs parked in the parking lot at each time period, the available reserve capacity varies. Therefore, the operator must consider arrival and departure times to make better decisions about the times that they want to sell the stored energy [11, 12].

All in all, it can be stated that the objective function is maximizing the benefit of parking lot. Applying the responsive demand model to the EV parking lot's features, the function can be expressed by a bi-level stochastic programming [11].

11.7.3 Application of IoT in Demand Response Programs Based on Parking Lots

As mentioned earlier, the available energy in EVs' batteries is a requirement asked by EV owners. These expectations can be stated as desired SOC of EVs at departure time. Therefore, the operator must predict the departure time of EVs' decreased uncertainty [13]. Thanks to IoT-enabled technologies, which are explained in Section 11.3 of this chapter, in addition to predicting the departure time based on historical pattern of EVs' behavior, it is possible for EV owners to send a signal through IoT communication channel to notify the parking lot operator about their decision some minutes prior to their actual departure. For example, people parks their EV at a parking lot to go to their workplace. After finishing work hours, they can send a signal to the parking lot operator that they have departed their workplace and will arrive at the parking lot in 20 minutes. Even, using smart device such as smartphones and smart watches, these devices can automatically send similar signal when EV owners reach in the parking lot zone (for example, when the distance between owner and the parking lot is less than a kilometer).

11.8 Charging Scheme for EVs Using IoT

Due to the increasing demand for energy and environmental problems caused by producing energy from fossil fuels, governments have allocated a huge amount of budget for developing new methods to generate energy with no or low environmental side effects. Among the suggested method, substituting EVs is one of the most promising methods, which can ameliorate the current condition. By this approach reducing fuel by internal combustion cars is warranted. This is because they do not use fossil fuels and the electricity can be generated by resources other than fossil fuels. Moreover, the efficiency of EVs is higher than conventional vehicles. In addition, energy conversion ratio at large power plants is much greater than conversion ratio at a car vehicle. Therefore, producing less amount of air pollutant is guaranteed. However, there are some challenges in this way. Increasing the number of EVs results in more electricity demand for charging EVs' batteries. This emerging electrical demand imposes a huge additional load to the power system. This additional demand results in new peak demands. Only 20% penetration of EVs could increment the peak demand up to 35%. Obviously, this amount of extra demand is not negligible. Thus, accurate energy management system must be employed for EVs' charging planning. This essential energy management system is required to shift EVs' charging demand to low demand periods, which is known as valley filling. Some well-known strategies, which are commonly used in energy management system, are price incentives and scheduling. To improve the effectiveness of these approaches, automated management systems must be deployed. For this sake, in addition to a well-developed system, sufficient information is required. This asks for highly reliable communication technology for data cognizance, aggregation, and processing. IoT as a technology that is developed recently for being used in smart cities can be used for this sake. The IoT potential for ubiquitous understanding of the various objects and real-time interaction with them permits this technology to be employed in the aforementioned energy management system. In the following, we will explain how an effective approach for energy management to accomplish charging scheduling using IoT-based system can be implemented.

It worth mentioning that, in general, two strategies can be develop: centralized and decentralized charging schemes [14]. Since the decentralized approach is easily scalable and is applicable for various problems from small cases to very large ones, this method is highly popular. Moreover, the whole complex problem can be divided into several smaller ones in decentralized approach, while in centralized strategy, the whole complicated problem must be solved by a single central computing unit. It is clear that a great computational burden is imposed to central computing unit at the centralized method. The problem gets worse when the amount of imported information increases. From robustness perspective, again decentralized method has some advantages. From privacy policy point of view, again, decentralized method is prior because each unit only has access to its own information. Therefore, the danger of information reveal is trivial in decentralized method, while in centralized strategy, information of all EVs must be sent to central unit, which endangers users' privacy. Due to the privileges of decentralized approach, many researchers have worked on developing this method. For example, Liasi and Bathaee have provided a game theory-based method to fill the load valley [15].

11.8.1 System Model and IoT Architecture

A general overview of a decentralized charging system for residential buildings is shown in Figure 11.10. Figure 11.10 depicts the EV charging infrastructure, which is based on IoT. As shown in Figure 11.10, a central energy management system is connected to several local management systems. Each local energy management system of ith parking lot is connected to N_i charging gateways. Each of them is employed to coordinate the EV charging of each household. The IoT architecture of EV utilizes data measuring equipment, which is integrated with the wireless

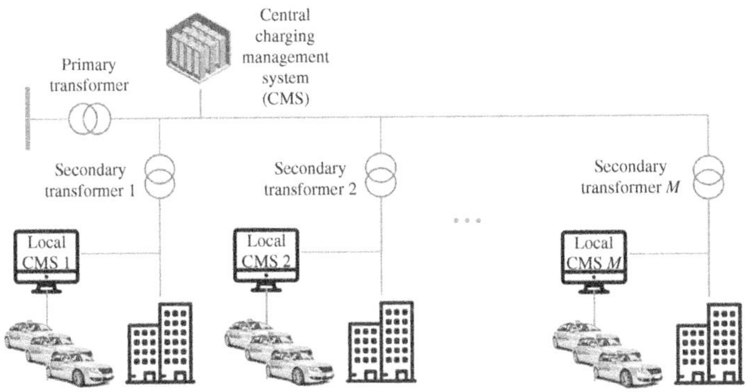

Figure 11.10 An overview of decentralized charging system for several residential buildings.

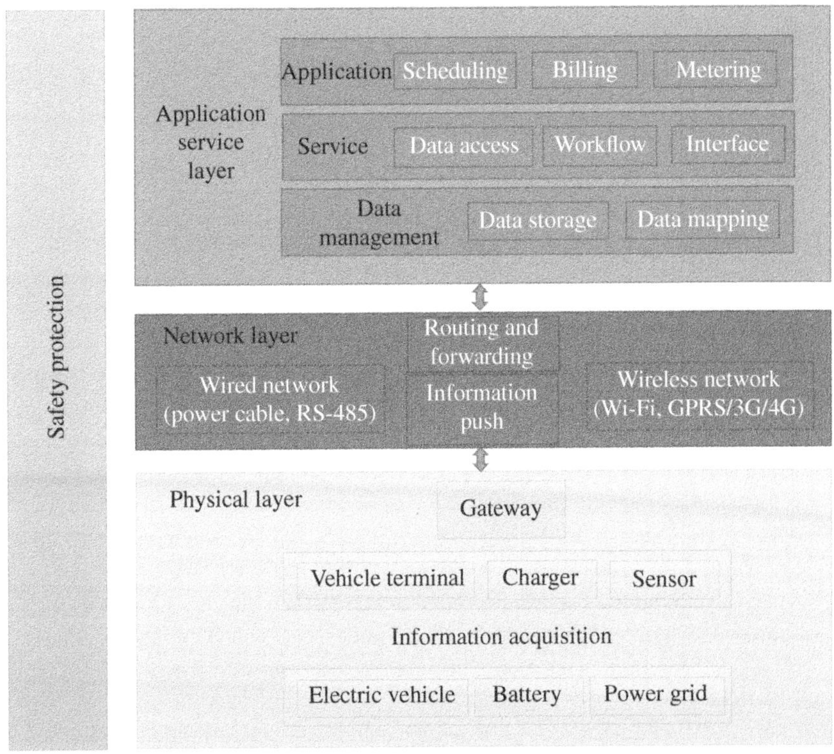

Figure 11.11 IoT architecture for EV charging system.

communication to obtain data, transmit it, and perform the EV charging scheduling. As shown in Figure 11.11, this architecture includes three layers [14]:

- **Physical layer**: The sensors, RFID tags, and measuring equipment are located at this layer. The information collected by the equipment at these level consists of battery SOC, EV ID, EV ON and OFF times, traveled distance, etc. For example, the charging gateway may obtain the EV SoC prior to initiation of charging procedure. Then, the cognizant data is transmitted through the charging gateways to the next layer.
- **Network layer**: Both wired (for example, RS-485) and wireless communications (for example, Wi-Fi or 3G/4G networks) are employed for completing the recognizing data transmission between various elements of charging network. The key issue at this layer is the information routing and active data push strategy.
- **Application service layer**: All recognized data attained at the network layer are saved and analyzed at this layer. For example, SOC and other parameters sensed at the physical layer are sent to central unit. This data is stored at the data base. Meanwhile, charging scheduling will be processed, and the latest commands can be sent to every charging gateway through Ethernet.

References

1 Vaidya, B. and Mouftah, H.T. (2019). IoT applications and services for connected and autonomous electric vehicles. *Arabian Journal for Science and Engineering* 45: 1–11.

2 SAE J3016 Standard (2021). Taxonomy and Definitions for Terms Related to Driving Automation Systems for On-Road Motor Vehicles J3016_202104. https://www.sae.org/standards/content/j3016_202104/ (accessed 16 November 2021).

3 Loeb, B., Kockelman, K.M., and Liu, J. (2018). Shared autonomous electric vehicle (SAEV) operations across the Austin, Texas network with charging infrastructure decisions. *Transportation Research Part C: Emerging Technologies* 89: 222–233.

4 Khalid, A., Sundararajan, A., Hernandez, A., and Sarwat, A.I. (2019). Facts approach to address cybersecurity issues in electric vehicle battery systems. In: *2019 IEEE Technology & Engineering Management Conference (TEMSCON)*, 1–6. IEEE.

5 Chan, A.C.-F. and Zhou, J. (2014). Cyber-physical device authentication for the smart grid electric vehicle ecosystem. *IEEE Journal on Selected Areas in Communications* 32 (7): 1509–1517.

6 Fraiji, Y., Azzouz, L.B., Trojet, W., and Saidane, L.A. (2018). Cyber security issues of internet of electric vehicles. In: *2018 IEEE Wireless Communications and Networking Conference (WCNC)*, 1–6. IEEE.
7 Rana, M.M. and Bo, R. (2020). IoT-based cyber-physical communication architecture: challenges and research directions. *IET Cyber-Physical Systems: Theory & Applications* 5 (1): 25–30.
8 Rana, M.M. (2019). IoT-based electric vehicle state estimation and control algorithms under cyber attacks. *IEEE Internet of Things Journal* 7 (2): 874–881.
9 Raghavan, S.S. (2016). Impact of demand response on electric vehicle charging and day ahead market operations. In: *2016 IEEE Power and Energy Conference at Illinois (PECI)*, 1–7. IEEE.
10 Liasi, S.G. and Golkar, M.A. (2017). Electric vehicles connection to microgrid effects on peak demand with and without demand response. In: *2017 Iranian Conference on Electrical Engineering (ICEE)*, 1272–1277. IEEE.
11 Shafie-khah, M., Heydarian-Forushani, E., Osório, G.J. et al. (2015). Optimal behavior of electric vehicle parking lots as demand response aggregation agents. *IEEE Transactions on Smart Grid* 7 (6): 2654–2665.
12 Abapour, M. and Zare, K. (2019). Stackelberg based optimal planning of DGs and electric vehicle parking lot by implementing demand response program. *Sustainable Cities and Society* 51: 101743.
13 Shojaabadi, S., Abapour, S., Abapour, M., and Nahavandi, A. (2016). Optimal planning of plug-in hybrid electric vehicle charging station in distribution network considering demand response programs and uncertainties. *IET Generation, Transmission and Distribution* 10 (13): 3330–3340.
14 Yao, L., Chen, Y.-Q., and Lim, W.H. (2015). Internet of things for electric vehicle: an improved decentralized charging scheme. In: *2015 IEEE International Conference on Data Science and Data Intensive Systems*, 651–658. IEEE.
15 Liasi, S.G. and Bathaee, S.M.T. (2017). Optimizing microgrid using demand response and electric vehicles connection to microgrid. In: *2017 Smart Grid Conference (SGC)*, 1–7. IEEE.

Author Index

Cyberphysical Smart Cities Infrastructures: Optimal Operation and Intelligent Decision Making, First Edition. Edited by M. Hadi Amini and Miadreza Shafie-khah.

d

e

f

g

h

i

j

k

l

m

S

t

u

v

Printed and bound by CPI Group (UK) Ltd, Croydon, CR0 4YY